BEHOLD THE SECOND
HORSEMAN

The Four Horsemen of the Apocalypse
Albrecht Durer, 1498

Compiled by Joseph B. Lumpkin and
Joyce A. Dujardin

BEHOLD THE SECOND HORSEMAN

For information about first time authors, contact Fifth Estate, Post Office Box 116, Blountsville, AL 35031.

First Edition

Cover Designed by An Quigley
Printed on acid-free paper

Library of Congress Control No: 2005925384

ISBN: 0-9760992-8-4

Fifth Estate, 2005

Dedication

They call it by a single name, but the beast has many heads and a thousand faces. It is the face of greed, the face of arrogance, the face of tyranny, of destruction, fear, aggression, and domination. Those whom it touches are destroyed and all who embrace it are devoured.

Matthew 24, 6-7. And ye shall hear of wars and rumors of wars: see that ye be not troubled: for all these things must come to pass, but the end is not yet. For nation shall rise against nation, and kingdom against kingdom: and there shall be famines, and pestilences, and earthquakes, in diverse places.

This book is dedicated to those who have fought for us, laying down life and limb to save us from ourselves. They stand guard between the best and worst of mankind; both of which reside within each of us, equally.

~ The Editors

Table of Contents

Prelude to War

THE UNIVERSAL SOLDIER
Donovan, 1968

He's five foot-two, and he's six feet-four,
He fights with missiles and with spears.
He's all of thirty-one, and he's only seventeen,
Been a soldier for a thousand years.

He's a Catholic, a Hindu, an Atheist, a Jain,
A Buddhist and a Baptist and a Jew.
And he knows he shouldn't kill,
And he knows he always will,
Kill you for me, my friend, and me for you.

And he's fighting for Democracy,
He's fighting for the Reds,
He says it's for the peace of all.
He's the one who must decide,
Who's to live and who's to die,
And he never sees the writing on the wall.

He's the Universal Soldier and he really is to blame,
His orders come from far away no more,
They come from here and there and you and me,
And brothers can't you see,
This is not the way we put the end to war.

Since the dawn of his existence, this creature, this homosapien, this fierce and violent beast has been obsessed with conflict. Like an incurable disease – an addictive drug, war has terrified him, teased and taunted him; repelled yet fascinated him. The power of conquest, the lure of sex, and the craving for wealth have fashioned the ultimate Beast of War – Man.

In his own words throughout the centuries, Man speaks to us of War – its allure and its tragedy. Come with us now and listen to the words of our own kind. Words that continue to haunt us in our nightmares.

- The Editors

And when he had opened the second seal, I heard the second beast say, Come and see. And there went out another horse that was red: and power was given to him that sat thereon to take peace from the earth, and that they should kill one another: and there was given unto him a great sword.
Holy Bible, Book of Revelation

I hate war as only a soldier who has lived it can, only as one who has seen its brutality, its stupidity.
General Dwight D. Eisenhower

Power corrupts, and absolute power corrupts absolutely.
Lord Acton

Victor and vanquished never unite in substantial agreement.
Tacitus

War is delightful to those who have no experience of it.
Erasmus

A bad peace is even worse than war.
Tacitus

Where force is necessary, there it must be applied boldly, decisively and completely. But one must know the limitations of force; one must know when to blend force with a maneuver, a blow with an agreement.
Leon Trotsky

The human race's prospects of survival were considerably better when we were defenseless against tigers than they are today when we have become defenseless against ourselves.
Arnold J. Toynbee

We are not interested in the possibilities of defeat. They do not exist.
Queen Victoria

The body of a dead enemy always smells sweet.
Titus Flavius Vespasian

The strongest of all warriors are these two - Time and Patience.
Leo Tolstoy.

Forewarned, forearmed; being prepared is half the victory.
Miguel de Cervantes

The foolish and the dead never change their opinions.
Jack Russell Lowell

The power of commitment is wondrous and can transcend all other forces.
Tak Kubota

He that is slow to anger is better than the mighty; and he that ruleth his spirit than he that taketh a city.
Holy Bible, Book of Proverbs

Pretend to be weak, that he may grow arrogant.
Sun Tzu

Trouble is easily overcome before it starts...Deal with it before it happens. Set things in order before there is confusion.
Lao Tzu

There is a tide in the affairs of men which, taken at the flood, leads on to fortune; omitted, all voyage in their life is bound in shallows and miseries. We must take the current when it serves, or lose our ventures.
William Shakespeare

Men are your castles. Men are your walls. Sympathy is your ally. Enmity is your foe.
Takeda Shingen

The warrior is not the brute. War makes them look alike. Life separates them fully.
Joseph Lumpkin

I feel I have this great creativity and spiritual force within me that is greater than faith, greater than ambition, greater than confidence, greater then determination, greater than vision. It is all of these combined. ... It is like a strong emotion mixed with faith, but a lot stronger.
Bruce Lee

In all aspects of life, relationships form the basis of everything. In all things, think with one's starting point in man.
Nabeshima Naoshige

Always forgive your enemies: nothing annoys them so much.
Oscar Wilde

No dictator, no invader, can hold an imprisoned population by force of arms forever. There is no greater power in the universe than the need for freedom. Against that power, governments and tyrants and armies cannot stand. The Centauri learned this lesson once. We will teach it to them again. Though it takes a thousand years, we will be free.
J. Michael Straczynski

I would rather die fighting than fight dying.
Kevin Grant

We will either find a way or make one.
Hannibal

It is not because things are difficult that we do not dare, it is because we do not dare that things are difficult.
Seneca

Human beings, who are almost unique in having the ability to learn from the experience of others, are also remarkable for their apparent disinclination to do so.
Douglas Noel Adams

The possession of unlimited power will make a despot of almost any man. There is a possible Nero in the gentlest human creature that walks.
Thomas Bailey

Nobody can give you freedom.
Malcolm X

Sometime they'll give a war and nobody will come.
Carl Sandburg

An army made up of creatures of impulse would be only a mob.
Ralph W. Sockman

The Samurai Code:
I have no parents; I make the Heavens and the Earth my parents.
I have no home; I make the Tan T'ien my home.
I have no divine power; I make honesty my Divine Power.
I have no means; I make Docility my means.
I have no magic power; I make personality my Magic Power.
I have neither life nor death; I make A Um my Life and Death.

I have no body; I make Stoicism my Body.
I have no eyes; I make The Flash of Lightning my eyes.
I have no ears; I make Sensibility my Ears.
I have no limbs; I make Promptitude my Limbs.
I have no laws; I make Self-Protection my Laws.

I have no strategy; I make the Right to Kill and the Right to Restore Life my Strategy.
I have no designs; I make Seizing the Opportunity by the Forelock my Designs.
I have no miracles; I make Righteous Laws my Miracle.
I have no principles; I make Adaptability to all circumstances my Principle.
I have no tactics; I make Emptiness and Fullness my Tactics.

I have no talent; I make Ready Wit my Talent.
I have no friends; I make my Mind my Friend.
I have no enemy; I make Incautiousness my Enemy.

I have no armor; I make Benevolence my Armor.
I have no castle; I make Immovable Mind my Castle.
I have no sword; I make No Mind my Sword.

War is an instrument entirely inefficient toward redressing wrong; and multiplies, instead of indemnifying losses.
Thomas Jefferson

For it has been said so truthfully that it is the soldier, not the reporter, who has given us the freedom of the press. It is the soldier, not the poet, who has given us freedom of speech. It is the soldier, not the agitator, who has given us the freedom to protest. It is the soldier who salutes the flag, serves beneath the flag, whose coffin is draped by the flag, who gives that protester the freedom to abuse and burn that flag.
Zell Miller

The conquest of the earth, which mostly means the taking it away from those who have a different complexion or slightly flatter noses than ourselves, is not a pretty thing when you look into it too much.
Joseph Conrad

Great men rejoice in adversity, just as brave soldiers triumph in war.
Seneca

If you will not fight for the right when you can easily win without bloodshed, if you will not fight when victory will be sure and not so costly, you may come to the moment when you will have to fight with all the odds against you and only a precarious chance of survival. There may be a worse case. You may have to fight when there is no chance of victory, because it is better to perish than to live as slaves.
Sir Winston Churchill

Right is more precious than peace.
Woodrow Wilson

War is an ugly thing, but not the ugliest of things. The decayed and degraded state of moral and patriotic feeling, which thinks that nothing is worth war is much worse. The person who has nothing for which he is willing to fight, nothing, which is more important than his own personal safety, is a miserable creature and has no chance of being free unless made and kept so by the exertions of better men than himself.
John Stuart Mill

All that is essential for the triumph of evil is that good men do nothing.
Edmund Burke

The God of War hates those who hesitate.
Euripides

Politics is war without bloodshed while war is politics with bloodshed.
Mao Tse Tung

Either war is obsolete or men are.
R. Buckminster Fuller

Never, never, never believe any war will be smooth and easy, or that anyone who embarks on the strange voyage can measure the tides and hurricanes he will encounter. The statesman who yields to war fever must realize that once the signal is given, he is no longer the master of policy but the slave of unforeseeable and uncontrollable events.
Sir Winston Churchill

The life of a modern soldier is ill represented by heroic fiction. War has means of destruction more formidable than the cannon and the sword.
Samuel Johnson

Chapter One – Prelude to War

Politics and War

OVER THERE
World War I Ballad by George M. Cohan, 1917

Over there, over there
Send the word, send the word over there
That the Yanks are coming, the Yanks are coming,

The drums rum-tumming ev'rywhere
So prepare say a pray'r
Send the word, send the word to beware
We'll be over, we're coming over,
And we won't come back till it's over over there!

If we choose, we may hear the drums and trumpets sounding their martial calls...the jingling of bridles on the massive warhorses...the deep rumblings of mechanized vehicles...the roar of lethal weapons from the sky. Dare we ask who sent them forward? Dare we question the commands of the Imperial They? In the deepness of our souls we have known Them and fear Them; They terrify us – the ones in the palaces, in the senates, in the parliaments...because we made Them.

- The Editors

We have met the enemy... and he is us.
Pogo

It belongs to human nature to hate those you have injured. It is only necessary to make war with five things: with the maladies of the body, with the ignorances of the mind, with the passions of the body, with the seditions of the city, with the discords of families.
Tacitus

Diplomats are just as essential to starting a war as soldiers are for finishing it. You take diplomacy out of war, and the thing would fall flat in a week.
Will Rogers

The time comes upon every public man when it is best for him to keep his lips closed.
Abraham Lincoln

A world without nuclear weapons would be less stable and more dangerous for all of us.
Margaret Thatcher

Of course it's the same old story. Truth usually is the same old story.
Margaret Thatcher

Vietnam presumably taught us that the United States could not serve as the world's policeman; it should also have taught us the dangers of trying to be the world's midwife to democracy when the birth is scheduled to take place under conditions of guerrilla war.
Jeanne Kirkpatrick

You can make a throne of bayonets, but you can't sit on it for long.
Boris Yeltsin

Some of the critics viewed Vietnam as a morality play in which the wicked must be punished before the final curtain and where any attempt to salvage self-respect from the outcome compounded the wrong. I viewed it as a genuine tragedy. No one had a monopoly on anguish.
Henry Kissinger

Above all, Vietnam was a war that asked everything of a few and nothing of most in America.
Myra MacPherson

No event in American history is more misunderstood than the Vietnam War. It was mis-reported then, and it is mis-remembered now.
Richard M. Nixon

All the wrong people remember Vietnam. I think all the people who remember it should forget it, and all the people who forgot it should remember it.
Michael Herr

Television brought the brutality of war into the comfort of the living room. Vietnam was lost in the living rooms of America-- not on the battlefields of Vietnam.
Marshall McLuhan

Power is like being a lady... if you have to tell people you are, you aren't.
Margaret Thatcher

Let us understand: North Vietnam cannot defeat or humiliate the United States. Only Americans can do that.
Richard M. Nixon

It was sheer professionalism and inspiration and the fact that you really cannot have people marching into other people's territory and staying there.
Margaret Thatcher

This war has already stretched the generation gap so wide that it threatens to pull the country apart.
Frank Church

You have a row of dominoes set up; you knock over the first one, and what will happen to the last one is that it will go over very quickly.
General Dwight D. Eisenhower

This is not a jungle war, but a struggle for freedom on every front of human activity.
Lyndon B. Johnson

We are at war with the most dangerous enemy that has ever faced mankind in his long climb from the swamp to the stars, and it has been said if we lose that war, and in so doing lose this way of freedom of ours, history will record with the greatest astonishment that those who had the most to lose did the least to prevent its happening.
Ronald Reagan

"Resource-constrained environment" are fancy Pentagon words that mean there isn't enough money to go around.
John W. Vessey, Jr.

A fanatic is one who can't change his mind and won't change the subject.
Sir Winston Churchill

You don't have a peaceful revolution. You don't have a turn-the-other-cheek revolution. There's no such thing as a nonviolent revolution. Revolution is bloody. Revolution is hostile. Revolution knows no compromise. Revolution overturns and destroys everything that gets in its way.
Malcolm X

I have witnessed the tremendous energy of the masses. On this foundation it is possible to accomplish any task whatsoever.
Mao Tse Tung

Standing in the middle of the road is very dangerous; you get knocked down by the traffic from both sides.
Margaret Thatcher

If you want to know the taste of a pear, you must change the pear by eating it yourself. If you want to know the theory and methods of revolution, you must take part in revolution. All genuine knowledge originates in direct experience.
Mao Tse Tung

It is a melancholy fact that many of the worst laws put upon the statute books have been put there with the best of intentions by thoroughly well meaning people. Mere desire to do right can no more by itself make a good statesman than it can make a good general.
Theodore Roosevelt

I do not want the best to be any more the deadly enemy of the good. We climb through degrees of comparison.
Archbishop Edward White Benson

Generally speaking, people do not care who is in charge as long as things run well. The charisma and expertise of the

leader should be used to build the organization, but things should not be tied so closely to the leader that his downfall would adversely impact the organization.
Joseph Lumpkin

You may have to fight a battle more than once to win it.
Margaret Thatcher

When you engage in actual fighting, if victory is long in coming, then men's weapons will grow dull and their ardor will be damped. If you lay siege to a town, you will exhaust your strength. Again, if the campaign is protracted, the resources of the State will not be equal to the strain. Thus, though we have heard of stupid haste in war, cleverness has never been seen associated with long delays. There is no instance of a country having benefited from prolonged warfare. In war, then, let your great object be victory, not lengthy campaigns. Thus it may be known that the leader of armies is the arbiter of the people's fate, the man on whom it depends whether the nation shall be in peace or in peril.
Sun Tzu

I have come to the conclusion that politics are too serious a matter to be left to the politicians.
General Charles DeGaulle

Politics is the gentle art of getting votes from the poor and campaign funds from the rich by promising to protect each from the other.
Oscar Ameringer

Patriotism is your conviction that this country is superior to all others because you were born in it.
George Bernhard Shaw

I'm extraordinarily patient provided I get my own way in the end.

Margaret Thatcher

A diplomat is someone who can tell you to go to hell in such a way that you look forward to the trip.
Cashie Stinnett

The real problem is what to do with the problem solvers after the problems are solved.
G. Talese

If you can't convince them, confuse them.
Harry S. Truman

The empires of the future will be the empires of the mind.
Sir Winston Churchill

In politics there is no honor.
Benjamin Disraeli

Few things are as immutable as the addiction of political groups to the ideas by which they have once won office.
John Kenneth Galbraith

Since a politician never believes what he says, he is surprised when others believe him.
General Charles DeGaulle

When a politician changes his position it's sometimes hard to tell whether he has seen the light or felt the heat.
Robert Fuoss

As a general rule the most successful man in life is the man who has the best information.
Benjamin Disraeli

In war, truth is so precious she must always be escorted by a bodyguard of lies.
Sir Winston Churchill

Among the calamities of war may be justly numbered the diminution of the love of truth, by the falsehoods which interest dictates and credulity encourages.
Samuel Johnson

Justice is what you get when you run out of money.
H.L. Mencken

Never in the field of human suffering, was so much made, by so few, from so many.
Sir Murray Rivers QC (Bryan Dawe)

An appeaser is one who feeds a crocodile - hoping that it will eat him last.
Sir Winston Churchill

It is better to have a lion at the head of an army of sheep than a sheep at the head of an army of lions.
Daniel Defoe

You cannot shake hands with a clenched fist.
Indira Gandhi

If error is corrected whenever it is recognized as such, the path of error is the path of truth.
Hans Reichenbach

I know that my unity with all people cannot be destroyed by national boundaries and government orders.
Leo Tolstoy

Where the willingness is great, the difficulties cannot be great.
Niccolò Machiavelli

In the councils of government, we must guard against the acquisition of unwarranted influence, whether sought or unsought, by the military-industrial complex. The potential for the disastrous rise of misplaced power exists and will persist.
General Dwight D. Eisenhower

Because just as good morals, if they are to be maintained, have need of the laws, so the laws, if they are to be observed, have need of good morals.
Niccolò Machiavelli

The greatest blunders, like the thickest ropes, are often compounded of a multitude of strands. Take the rope apart, separate it into the small threads that compose it, and you can break them one by one. You think, 'That is all there was!' But twist them all together and you have something tremendous.
Victor Hugo

I never did give anybody hell. I just told the truth, and they thought it was hell.
Harry S Truman

How can you govern a country which has 246 varieties of cheese?
General Charles DeGaulle

We live in a Newtonian world of Einsteinian physics ruled by Frankenstein logic.
David Russell

A single death is a tragedy; a million deaths is a statistic.
Joseph Stalin

I am not bound to win, but I am bound to be true. I am not bound to succeed, but I am bound to live by the light that I have. I must stand with anybody that stands right, and stand with him while he is right, and part with him when he goes wrong.
Abraham Lincoln

I can make more generals, but horses cost money.
Abraham Lincoln

It is often easier to fight for one's principles than to live up to them.
Adlai Stevenson

If you do not tell the truth about yourself, you cannot tell it about other people.
Virginia Woolf

To err is human, but when the eraser wears out ahead of the pencil, you're overdoing it.
Josh Jenkins

Life can only be understood backwards; but it must be lived forwards.
Soren Kierkegaard

If it ain't broke, break it.
Richard Pascale

Oppose, adapt, adopt.
Benjamin Disraeli

There can be no justice so long as rules are absolute.
Patrick Stewart

The great masses of people will more easily fall victims to a big lie than to a small one, especially if it is repeated over and over.
Adolph Hitler

The lie can be maintained only for such time as the State can shield the people from the political, economic and/or military consequences of the lie. It thus becomes vitally important for the State to use all of its powers to repress dissent, for the truth is the mortal enemy of the lie, and thus by extension, the truth becomes the greatest enemy of the State.
Dr. Joseph M. Goebbels

In our time, political speech and writing are largely the defense of the indefensible.
George Orwell

I know what a statesman is. He's a dead politician. We need more statesmen.
Robert C. Edwards

Few people think more than two or three times a year; I have made an international reputation for myself by thinking once or twice a week.
George Bernard Shaw

Bad administration, to be sure, can destroy good policy; but good administration can never save bad policy.
Adlai Stevenson

HISTORY, n. An account mostly false, of events mostly unimportant, which are brought about by rulers, mostly knaves, and soldiers, mostly fools.
Ambrose Bierce

I like to believe that people in the long run are going to do more to promote peace than our governments. Indeed, I think that people want peace so much that one of these days governments had better get out of their way and let them have it.
General Dwight D. Eisenhower

The more corrupt the republic, the more numerous the laws.
Tacitus

There is nothing that surpasses ruling with benevolence. However, to put into practice enough benevolent governing to rule the country is difficult. To do this lukewarmly will result in neglect. If governing with benevolence is difficult, then it is best to govern strictly. To govern strictly means to be strict before things have arisen, and to do things in such a way that evil will not arise. To be strict after the evil has arisen is like laying a snare. There are few people who will make mistakes with fire after having once been burned. Of people who regard water lightly, many have been drowned.
Tzu Ch'an

If you are able to vote, then do so. There may be no candidates or issues you want to vote for... but there will certainly be someone or something to vote against. In case of doubt, vote against. By this rule you will rarely go wrong.
Lazarus Long

In time of war the first casualty is truth.
Boake Carter

It is patriotic to support him insofar as he efficiently serves the country. It is unpatriotic not to oppose him to the exact extent that by inefficiency or otherwise he fails in his duty to stand by the country.
Theodore Roosevelt

Are you a politician who says to himself: I will use my country for my own benefit? Or are you a devoted patriot, who whispers in the ear of his inner self: I love to serve my country as a faithful servant?
Kahlil Gibran

Each man must for himself alone decide what is right and what is wrong, which course is patriotic and which isn't. You cannot shirk this and be a man.
Thomas Tusser

In war, as in life, it is often necessary, when some cherished scheme has failed, to take up the best alternative open, and if so, it is folly not to work for it with all your might.
Sir Winston Churchill

People never lie so much as after a hunt, during a war or before an election.
Otto von Bismarck

The whole aim of practical politics is to keep the populace alarmed (and hence clamorous to be led to safety) by menacing it with an endless series of hobgoblins, all of them imaginary.
H.L. Mencken

War is mainly a catalogue of blunders.
Sir Winston Churchill

Conceit, arrogance and egotism are the essentials of patriotism. Patriotism assumes that our globe is divided into little spots, each one surrounded by an iron gate. Those who had the fortune of being born on some particular spot, consider themselves better, nobler, grander, more intelligent than the living beings inhabiting any other spot. It is, therefore, the duty of everyone living on that chosen spot to fight, kill, and die in the attempt to impose his superiority upon all others.
Emma Goldman

When a whole nation is roaring patriotism at the top of its voice, I am fain to explore the cleanness of its hands and the purity of its heart.
Ralph Waldo Emerson

Herein lies a riddle: How can a people so gifted by God become so seduced by naked power, so greedy for money, so addicted to violence, so slavish before mediocre and treacherous leadership, so paranoid, deluded, lunatic?
Philip Berrigan

If the Nuremberg laws were applied, then every post-war American president would have been hanged.
Noam Chomsky

Praising our leaders, we're getting in tune with the music played by the madmen.
Alphaville

Authoritarian government required to speak, is silent. Representative government required to speak, lies with impunity.
Napoleon Bonaparte

It is the duty of the patriot to protect his country from its government.
Thomas Paine

Few of us can easily surrender our belief that society must somehow make sense. The thought that The State has lost its mind and is punishing so many innocent people is intolerable. And so the evidence has to be internally denied.
Arthur Miller

There exists a shadowy Government with its own Air Force, its own Navy, its own fundraising mechanism, and the ability to pursue its own ideas of national interest, free from all checks and balances, and free from the law itself.
Senator Daniel K. Inouye

The people can have anything they want. The trouble is, they do not want anything. At least they vote that way on election day.
Eugene Debs

Why of course the people don't want war. Why should some poor slob on a farm want to risk his life in a war when the best he can get out of it is to come back to his farm in one piece? Naturally, the common people don't want war: neither in Russia, nor in England, nor for that matter in Germany. That is understood. But after all it is the leaders of the country who determine the policy, and it is always a simple matter to drag the people along, whether it is a democracy, or a fascist dictatorship, or a parliament, or a communist dictatorship. Voice or no voice, the people can always be brought to the bidding of the leaders. That is easy. All you have to do is to tell them they are being attacked, and denounce the pacifists for lack of patriotism and exposing the country to danger.
Hermann Goering

Truth will do well enough if left to shift for herself. She has no need of force to procure entrance into the minds of men.
Thomas Jefferson

It does not require a majority to prevail, but rather an irate, tireless minority keen to set brush fires in people's minds.
Samuel Adams

Never give in, never, never, never, never, in nothing great or small, large or petty, never give in except to convictions of honor and good sense. Never yield to force ... never yield to the apparently overwhelming might of the enemy.
Sir Winston Churchill

We shall fight on the beaches. We shall fight on the landing grounds. We shall fight in the fields, and in the streets, we shall fight in the hills. We shall never surrender!
Sir Winston Churchill

History does not long entrust the care of freedom to the weak or timid.
General Dwight D. Eisenhower

No man is entitled to the blessings of freedom unless he be vigilant in its preservation.
General Douglas MacArthur

The political object is the goal, war is the means of reaching it, and the means can never be considered in isolation from their purposes.
Carl von Clausewitz

The only power tyrants have, is the power relinquished to them by their victims.
Ettiene de la Boetie

With reasonable men I will reason; with humane men I will plead; but to tyrants I will give no quarter, nor waste arguments, where they will certainly be lost.
William Lloyd Garrison

War is a game that is played with a smile. If you can't smile, grin. If you can't grin, keep out of the way till you can.
Sir Winston Churchill

One should never allow chaos to develop in order to avoid going to war, because one does not avoid a war but instead puts it off to his disadvantage.
Niccolò Machiavelli

Laws are inoperative in war.
Marcus Tullius Cicero

War is the continuation of politics by other means.
Carl von Clausewitz

Controlled, universal disarmament is the imperative of our time. The demand for it by the hundreds of millions whose chief concern is the long future of themselves and their children will, I hope, become so universal and so insistent that no man, no government anywhere, can withstand it.
General Dwight D. Eisenhower

The ballot is stronger than the bullet.
Abraham Lincoln

Business and War

JOHNNY HAS GONE FOR A SOLDIER
A Revolutionary War Song

O Johnny dear has gone away
He has gone afar across the bay,

O my heart is sad and weary today,
Johnny has gone for a soldier.

In the boardrooms and the battlefields, in flying machines at 30,000 feet...the art of war has always been the same.

The weapons are equally deadly. The rule of no rule applies. Strength, power, poise, deception, and the luck of the moment determine the outcome of the battle...and the victorious always write history.

The Editors

But war is not the whole business of life; it happens but seldom, and every man, either good or wise, wishes that its frequency were still less. That conduct which betrays designs of future hostility, if it does not excite violence, will always generate malignity; it must forever exclude confidence and friendship, and continue a cold and sluggish rivalry, by a sly reciprocation of indirect injuries, without the bravery of war or the security of peace.
Samuel Johnson

Motivation is everything. You can do the work of two people, but you can't be two people. Instead, you have to inspire the next guy down the line and get him to inspire his people.
Lee Iacocca

If you want to succeed, you should strike out on new paths rather than travel the worn paths of accepted success.
John D. Rockefeller, Jr.

All the business of war, and indeed all the business of life, is to endeavor to find out what you don't know by what you do; that's what I called 'guess what was at the other side of the hill'.
Duke of Wellington

Eagles don't flock -- You have to find them one at a time.
H Ross Perot

If you are going to try to go to war, or to prepare for war, in a capitalist country, you have got to let business make money out of the process or business won't work.
Henry Lewis Stimson

Making a killing – the business of war.
Phillip Van Niekerk

If you have ideas, you have the main asset you need, and there isn't any limit to what you can do with your business and your life. They are any man's greatest asset -- IDEAS.
Harvey S. Firestone

Nothing is illegal if a hundred businessmen decide to do it, and that's true anywhere in the world.
Andrew Young

So long as war is the main business of nations, temporary despotism --despotism during the campaign --is indispensable.
Walter Bagehot

When two men in business always agree, one of them is unnecessary.
W. Wrigley Jr.

The only competition worthy of a wise man is with himself.
Washington Allston

You can get much farther with a kind word and a gun than you can with a kind word alone.
Al Capone

The secret of success is to know something nobody else knows.
Aristotle Onassis

You can't do business sitting on your arse.
Lord MacLaurin

The rule of business is how fast you can get your idea to market. Those whose systems do not allow them to move quickly are doomed.
Ken Tuchman

The art of giving advice is to make the recipient believe he thought of it himself.
Frank Tyger

Dreaming is zero value. I mean, anyone can dream.
Bill Gates

He who never fell, never climbed.
Anonymous

I dream of a company where people come to work every day in a rush to try something they woke up thinking about the night before. We want them to go home from work wanting to talk about what they did that day, rather than trying to forget it. We want factories where the whistle blows and everybody wonders where the time went, and then somebody suddenly wonders aloud why we need a whistle. We want a company where people find a better way, everyday, of doing things, and where by shaping their own work experience, they make their lives better and their company the best.
Jack Welch

Make three correct guesses consecutively and you will establish a reputation as an expert.
Laurence J. Peter

Success depends on your backbone, not your wishbone.
Anonymous

Obstacles are things a person sees when he takes his eyes off his goal.
E. Joseph Cossman

When there is an original sound in the world, it makes a hundred echoes.
John Shedd

In any great organization it is far, far safer to be wrong with the majority than to be right alone.
John Kenneth Galbraith

There are two kinds of people, those who do the work and those who take the credit. Try to be in the first group; there is less competition there.
Indira Gandhi

Education's purpose is to replace an empty mind with an open one.
Malcolm S. Forbes

It is not the employer who pays the wages. Employers only handle the money. It is the customer who pays the wages.
Henry Ford

The test of a first-rate intelligence is the ability to hold two opposing ideas in mind at the same time and still retain the ability to function.
F. Scott Fitzgerald

Risk-taking is the essence of innovation.
Herman Kahn

Better ask questions twice than lose your way once.
Danish Proverb

People who are just in it for the money - they usually fail.
Robert Holmes A Court.

Thinking is the hardest work there is, which is probably the reason why so few engage in it.
Henry Ford

If you're not confused about the current state of the economy then you clearly do not understand what is going on.
Dr Chris Caton

Deciding what to do is easy, deciding what not to do is hard.
Michael Dell

To every man, every day, will come one valuable thought.
Thomas Edison

Systems allow us to apply the best thinking and give us a benchmark against which to measure and evaluate future ideas.
Alan Patching

Imagination is more important than knowledge.
Albert Einstein

Remarkable people in all fields of endeavor move the world forward - they never give up.
Kevin Gosper

Once you have passionately sold the core values of your organization to your people, they have a very simple choice, and that is to be an Ambassador or an Assassin of those values.
Geoff Burch

Only a radically new kind of creativity will keep you and your organization up there with the best.
Dr Kobus Neethling

Success is the maximum utilization of the ability that you have.
Zig Zilglar

Everyone is a genius at least once a year; a real genius has his original ideas closer together.
Georg Lichtenberg

I have not failed. I've just found 10,000 ways that won't work.
Thomas Alva Edison

Not everything that can be counted counts, and not everything that counts can be counted.
Albert Einstein

The race for market share is a race against time, not against competitors. Good ideas sooner always beat good ideas eventually. Today's marketplace is more concerned about when than who.
Bruce Haddon

Successful is the person who has lived well, laughed often and loved much, who has gained the respect of children, who leaves the world better than they found it, who has never lacked appreciation for the earth's beauty, who never fails to look for the best in others or give the best of themselves.
Ralph Waldo Emerson

Age is only a number, a cipher for the records. A man can't retire his experience. He must use it. Experience achieves more with less energy and time.
Bernard Baruch

Whenever you see a successful business, someone once made a courageous decision.
Peter F. Drucker

When you reduce a complex message to something customers can understand, you're not dumbing it down. You're smartening it up.
Bruce Haddon

If you haven't had a failure in this business, you haven't been around long enough.
Sandra Levy

In a competitive world so much importance is put on winning at any cost that we sometimes forget that honesty, decency and integrity are the ultimate victors in both business and life.
Bryce Courtenay

To manage a business successfully requires as much courage as that possessed by the soldier who goes to war. Business courage is the more natural because all the benefits which the public has in material wealth come from it.
Charles F. Abbott

I believe that banking institutions are more dangerous to our liberties than standing armies if the American people ever allow private banks to control the issue of currency...the banks and corporations that will grow up around them will deprive the people of all property until their children will wake up homeless on the continent that their fathers conquered.
Thomas Jefferson

Such as it is, the press has become the greatest power within the Western World, more powerful than the legislature, the executive and judiciary. One would like to ask: by whom has it been elected, and to whom is it responsible?
Alexander Solzhenitsyn

Do not fear the enemy, for your enemy can only take your life. It is far better that you fear the media, for they will steal your HONOR. That awful power, the public opinion of a nation, is created in America by a horde of ignorant, self-complacent simpletons who failed at ditching and shoemaking and fetched up in journalism on their way to the poorhouse.
MarkTwain

Business is war. It is a battle to reach your objectives. It is a battle to keep your key talent from leaving and "upgrading" your competitors. Above all, it is a battle to dominate your competition.
Greg Langston

Business is war...war is business.
Carl von Clausewitz

The Philosophy of War

WE ARE TENTING TONIGHT
Civil War Ballad by Walter Kittredge, 1864

Many are the hearts that are weary tonight,
Wishing for the war to cease;

Many are the hearts looking for the right
To see the dawn of peace.

Tenting tonight, Tenting tonight,

Tenting on the old Campground.

The most brilliant minds, the most profound thinkers demand answers to the indefinable quality and the finality of Man's existence. They seek, and as the millennia pass, they still seek. They seek the answers in war and in peace ... yet they cannot begin to conceive the questions which must be asked.

- The Editors.

Great warrior! Wars not make one great!
Yoda, The Empire Strikes Back

Four things come not back:
The spoken word,
The sped arrow,
The past life,
The neglected opportunity.
Arabian Proverb

I have a dream that one day this nation will rise up and live out
the true meaning of its creed: 'We hold these truths to be self-
evident that all men are created equal'.
Martin Luther King Jr.

A house divided against itself cannot stand. I believe this
government cannot endure permanently half slave and half
free.
Abraham Lincoln

Europe was created by history. America was created by
philosophy.
Margaret Thatcher

The end may justify the means as long as there is something
that justifies the end.
Leon Trotsky

The historic ascent of humanity, taken as a whole, may be
summarized as a succession of victories of consciousness over
blind forces - in nature, in society, in man himself.
Leon Trotsky

There are no absolute rules of conduct, either in peace or war. Everything depends on circumstances.
Leon Trotsky

There is a limit to the application of democratic methods. You can inquire of all the passengers as to what type of car they like to ride in, but it is impossible to question them as to whether to apply the brakes when the train is at full speed and accident threatens.
Leon Trotsky

Believe that we too love freedom and desire it. To us it is more desirable than anything in the world. If you strike us down now, we shall rise again and renew the fight. You cannot conquer Ireland; you cannot extinguish the Irish passion for freedom; if our deed has not been sufficient to win freedom then our children will win it with a better deed.
Padraig Pearse

A handful of men, inured to war, proceed to certain victory, while on the contrary, numerous armies of raw and undisciplined troops are but multitudes of men dragged to the slaughter.
Flavius Renatus Vegetius

He who puts out his hand to stop the wheel of history will have his fingers crushed.
Lech Walesa

History is made at night. Character is what you are in the dark.
Lord John Whorfin

Classes struggle, some classes triumph, others are eliminated. Such is history; such is the history of civilization for thousands of years.
Mao Tse Tung

A man who has committed a mistake and doesn't correct it is committing another mistake
Confucius

When valor preys on reason, it eats the sword it fights with.
William Shakespeare

He who will not apply new remedies must expect old evils.
Sir Francis Bacon

If the teacher is not respected and the students not cared for, confusion will arise, however clever one is. This is the crux of mystery.
Lao Tzu

The principle on which to manage an army is to set up one standard of courage which all must reach.
Sun Tzu

All men come to he who keeps unity. For there lie rest, happiness, and peace.
Lao Tzu

The humble is the root of nobility. Low is the foundation of high. Princes and lords consider themselves orphaned, widowed, and worthless. Do they not depend on being humble? Too much success is not an advantage. Do not tinkle like jade or clatter like a stone chime.
Lao Tzu

And if we are able thus to attack an inferior force with a superior one, our opponents will be in dire straits.
Sun Tzu

That the impact of your army may be like a grindstone dashed against an egg--this is effected by the science of weak points and strong. In all fighting, the direct method may be used for joining battle, but indirect methods will be needed in order to secure victory.
Sun Tzu

Rapidity is the essence of war: take advantage of the enemy's unreadiness, make your way by unexpected routes, and attack unguarded spots.
Sun Tzu

Disciplined and calm, to await the appearance of disorder and hubbub amongst the enemy - this is the art of retaining self-possession.
Sun Tzu

Patience is not passive. Patience is concentrated strength.
Bruce Lee

One must not be negligent of learning. Lun Yu says, to study and not to think is darkness. To think without study is dangerous.
Takeda Nobushige

If you sit, sit. If you stand, stand but never wobble.
Master Ummon

Conquering evil, not the opponent, is the essence of swordsmanship.
Yagyu Munenori

In connection with military matters, one must never say what can absolutely not be done. By this, the limitations of one's heart will be exposed.
Asakura Norikage

Conquer the self and you will conquer the opponent.
Takuan Soho

Argue for your limitations, and sure enough, they are yours.
Richard Bach

It's a difficult thing to truly know your own limits and points of weakness.
From the Hagakure

He who is aware of his own weakness will remain master of himself in any situation.
Gichin Funakoshi

The mind of the warrior remains focused on his own mortality. The sacredness and brevity of life is always in his thoughts. Life is lived to the fullest, moment by moment, when the possibility of death is realized. The warrior's heart is not reserved for those who do battle with others, but is kept secret for those who battle themselves and their own limitations.
Joseph Lumpkin

Take the arrow in your forehead, but never in your back.
HwaRang maxim

There is no such thing as an effective segment of totality.
Bruce Lee

Whenever you meet difficult situations dash forward bravely and joyfully.
Tsunetomo Yamamoto

The man whose profession is arms (fighting) should calm his own spirit and look into the depths of others. Doing so is likely the best of the martial arts.
Shiba Yoshimasa

Yield and overcome. Bend and be straight. Empty and be filled. Wear out and become new. Have little and gain. Have much and be confused.
Lao Tsu

If a man becomes alienated from his friends, he should make endeavors in the way of humanity.... One should not turn his back on reproof.
Takeda Nobushige

Knowing your ignorance is strength. Ignoring knowledge is sickness. When one becomes sick of sickness he is no longer sick.
Lao Tzu

It is hardly necessary to record that both learning and the military arts are the Way of the Warrior, for it is an ancient law that one should have Learning on the left and martial arts on the right.
Hojo Nagauji

The martial arts consider intelligence most important because intelligence involves the ability to plan and to know when to change effectively.
Sun Tzu

The sage takes care of all men, and abandons no one. He takes care of all things and abandons nothing. This is called following the light.
Lao Tzu

I have heard that when a man has literary business, he will always take military preparations; and when he has military business, he will always take literary preparations.
Confucius

Without knowledge of learning, one will have no military victories.
Imagawa Sadayo

A goal is not always meant to be reached, it often serves simply as something to aim at.
Bruce Lee

Achieve results, but never glory in them. Achieve results, but do not boast. Achieve results, but do not be proud. Achieve results, because it is the natural way. Achieve results, but not through violence.
Lao Tzu

What others teach, I also teach; that is, A violent man will die a violent death. This is the essence of my teaching.
Lao Tsu

Train. An unpolished crystal does not shine; an undisciplined Samurai does not have brilliance. A Samurai therefore should cultivate his mind.
Anonymous

To invalidate the opponent's expectation, a person must know others and know the self.
Sun Tzu

Water shapes its course according to the nature of the ground over which it flows; the soldier works out his victory in relation to the foe who he is facing. Therefore, just as water retains no constant shape, so in warfare there are no constant conditions. He who can modify his tactics in relation to his opponent and thereby succeed in winning, may be called a heaven-born captain.
Sun Tzu

The art of war is to avoid big battles.
Sun Tzu

Experience is the name that everyone gives to his mistakes.
Oscar Wilde.

Being defeated is often a temporary condition. Giving up is what makes it permanent.
Marilyn Vos Savant

Education is when you read the fine print. Experience is what you get if you don't.
Pete Seeger

The breakfast of champions is not cereal, it's the opposition.
Nick Seitz

The brighter you are, the more you have to learn.
Don Herold

It is the eternal struggle between these two principles - right and wrong. They are the two principles that have stood face to face from the beginning of time and will ever continue to struggle. It is the same spirit that says, You work and toil and earn bread, and I'll eat it.
Abraham Lincoln

Do not do unto others as you expect they should do unto you. Their tastes may not be the same.
George Bernard Shaw

We are most nearly ourselves when we achieve the seriousness of the child at play.
Heraclitus

I am free of all prejudices. I hate everyone equally.
W.C. Fields

In the midst of great joy do not promise anyone anything. In the midst of great anger do not answer anyone's letter.
Chinese Proverb

A smooth sea never made a skillful mariner.
English Proverb

Consistency is the hobgoblin of small minds.
J Frank Dobie

Insanity in individuals is something rare - but in groups, parties, nations and epochs, it is the rule.
Friedrich Nietzsche

Everybody sets out to do something, and everybody does something, but no one does what he sets out to do.
George Moore

To achieve great things, we must live as though we are never going to die.
Luc de Clarnes Vauvenargues

Optimism is the faith that leads to achievement. Nothing can be done without hope.
Helen Keller

Happiness is not a possession to be prized. It is a quality of thought, a state of mind.
Daphne du Maurier

It is better to deserve honors and not have them, than to have them and not deserve them.
Mark Twain

Cherish your visions and your dreams as they are the children of your soul; the blueprints of your ultimate achievements.
Napoleon Hill

Glory is fleeting, but obscurity is forever.
Napoleon Bonaparte

Don't be so humble - you are not that great.
Golda Meir

People demand freedom of speech to make up for the freedom of thought which they avoid.
Soren Kierkegaard

Jokes of the proper kind, properly told, can do more to enlighten questions of politics, philosophy, and literature than any number of dull arguments.
Isaac Asimov

Doing easily what others find difficult is talent; doing what is impossible for talent is genius.
Henri-Frédéric Amiel

It is the mark of an educated mind to be able to entertain a thought without accepting it.
Aristotle

All things that are truly great are at first thought impossible.
Friedrich Nietzsche

Dare to believe only in yourself.
Friedrich Nietzsche

Hegel was right when he said that we learn from history that man can never learn anything from history.
George Bernard Shaw

It is hard enough to remember my opinions, without also remembering my reasons for them!
Friedrich Nietzsche

Man is the cruelest animal.
Friedrich Nietzsche

I would never die for my beliefs because I might be wrong.
Bertrand Russell

Science is what you know, philosophy is what you don't know.
Bertrand Russell

All knowledge, we feel, must be built up upon our instinctive beliefs; and if these are rejected, nothing is left.
Bertrand Russell

We think in generalities, but we live in detail.
Alfred North Whitehead

Any man may make a mistake; none but a fool will persist in it.
Marcus Tullius Cicero

Each morning puts a man on trial and each evening passes judgment.
Ray L. Smith

It is better to be defeated on principle than to win on lies.
Arthur Caldwell

How shall we rank thee upon glory's page, Thou more than soldier, and just less than sage?
Thomas Moore

The first who was king was a fortunate soldier: Who serves his country well has no need of ancestors.
Voltaire

What can they see in the longest kingly line in Europe, save that it runs back to a successful soldier?
Sir Walter Scott

Shall I ask the brave soldier who fights by my side in the cause of mankind, if our creeds agree?
Charles Lamb

The schoolmaster is abroad, and I trust to him, armed with his primer, against the soldier in full military array.
Henry Peter Brougham

Although too much of a soldier among sovereigns, no one could claim with better right to be a sovereign among soldiers.
Sir Walter Scott

In war, there are no unwounded soldiers.
Jose Narosky

The Way lies at hand yet it is sought afar off; the thing lies in the easy yet it is sought in the difficult.
Mencius

The dance of battle is always played to the same impatient rhythm. What begins in a surge of violent motion is always reduced to the perfectly still.
Sun Tzu

The undisturbed mind is like the calm body water reflecting the brilliance of the moon. Empty the mind and you will realize the undisturbed mind.
Yagyu Jubei

You might as well stand and fight because if you run, you will only die tired.
Vern Jocque - Sei Shin Kan.

Am I not destroying my enemies when I make friends of them?
Abraham Lincoln

It is easy to kill someone with a slash of a sword. It is hard to be impossible for others to cut down.
Yagyu Munenori

Mental bearing (calmness), not skill, is the sign of a matured samurai. A Samurai therefore should neither be pompous nor arrogant.
Sukahara Bokuden.

One finds life through conquering the fear of death within one's mind. Empty the mind of all forms of attachment, make a go-for-broke charge and conquer the opponent with one decisive slash.
Togo Shigekata.

It does not matter how slowly you go so long as you do not stop.
Confucius

I have a high art, I hurt with cruelty those who would damage me.
Archilocus

You must concentrate upon and consecrate yourself wholly to each day, as though a fire were raging in your hair.
Taisen Deshimaru

Given enough time, any man may master the physical. With enough knowledge, any man may become wise. It is the true warrior who can master both....and surpass the result.
Tien T'ai

Act like a man of thought - Think like a man of action.
Thomas Mann

Civilize the mind but make savage the body.
Mao Tse Tung

The belief in the possibility of a short decisive war appears to be one of the most ancient and dangerous of human illusions.
Robert Lynd

The violence of war admits no distinction; the lance, that is lifted at guilt and power, will sometimes fall on innocence and gentleness.
Samuel Johnson

I think that, as life is action and passion, it is required of a man that he should share the passion and action of his time at peril of being judged not to have lived.
Oliver Wendell Holmes Jr.

True knowledge is to experience the inner self, but since the inner being is unique to every individual, knowledge cannot be assimilated by talking about it.
Theun Mares

Knowledge gained from someone else lacks the confidence necessary to implement that knowledge. Confidence is cultivated only through practice.
Theun Mares

As we have already seen, at the end of the day it does not matter how many battles we have won or how many we have lost, as the only thing of importance is whether or not we fought and, if we did, how well we fought. Did we run from a battle because of fear, or did we fight bravely, giving it our all?
Theun Mares

Explanations are not reality – only a makeshift arrangement of the world.
Theun Mares

Confusion is a willfully induced state of mind. We can enter or exit it at will. Man deliberately confuses himself in order to plead ignorance.
Theun Mares

Confusion is a most convenient escapism used by man whenever he has to face something that frightens him, or that he does not like. However, we are always fully aware of what we are doing, even though we may choose not to acknowledge our true motives.
Theun Mares

The warrior, knowing that there is nothing to understand, acknowledges a barrier when he comes to it, and then jumps over it.
Theun Mares

When the warrior encounters a problem in his life he puts his mind at rest by acknowledging it for the obstacle it is, but instead of getting caught up in rationalizations in an effort to understand the problem, he simply tackles it immediately. Problems in themselves have no value other than to make us emotionally stronger, mentally more agile and spiritually wiser.
Theun Mares

Should a warrior feel the need to be comforted, he simply chooses anyone or anything, be it a friend, dog, or mountain, to whom he expresses his innermost feelings. It does not matter to the warrior if he is not answered, or if he is not heard, because the warrior not seeking to be understood or helped – by verbalizing his feelings, he is merely releasing the pressure of his battle.
Theun Mares

The hunter is intimately familiar with his world, yet remains detached from it.
Theun Mares

Ultimately, you must forget about technique. The further you progress, the fewer teachings there are. The Great Path is really NO PATH.
Ueshiba Morihei

In the beginners mind there are many possibilities, but in the expert's mind there are few.
Suzuki

To practice Zen or the Martial Arts, you must live intensely, wholeheartedly, without reserve - as if you might die in the next instant.
Taisen Deshimaru

Don't think dishonestly
The Way is in training
Become acquainted with every art
Know the ways of all professions
Distinguish between gain and loss
Develop intuitive judgment and understanding for everything
Perceive those things which cannot be seen
Pay attention even to trifles
Do nothing which is of no use.
Mayomoto Musashi

Empty your mind,
Be formless, shapeless, like water.
Now you put water into a cup, it becomes the cup.
You put water into a bottle, it becomes the bottle.
You put water into a teapot, it becomes the teapot.
Now water can flow, or it can crash,
Be water my friend.
Bruce Lee

The consciousness of self is the greatest hindrance to the proper execution of all physical action.
Bruce Lee

Victory goes to the one who has no thought of himself.
Shinkage School of Swordsmanship

It is truly regrettable that a person will treat a man who is valuable to him well, and a man who is worthless to him poorly.
Samurai Quotation

It is a principle of the art of war that one should simply lay down his life and strike. If one's opponent also does the same, it is an even match. Defeating one's opponent is then a matter of faith and destiny.
Yamamoto Tsunetomo

If a warrior is not unattached to life and death, he will be of no use whatsoever. The saying that All abilities come from one mind sounds as though it has to do with sentient matters, but it is in fact a matter of being unattached to life and death. With such non-attachment one can accomplish any feat. Martial arts and the like are related to this insofar as they can lead to the Way.
Yamamoto Tsunetomo

This is essentially a people's contest... whose leading object is to elevate the condition of men - to lift artificial weights from all shoulders - to clear the paths of laudable pursuit for all - to afford all, an unfettered start and a fair chance, in the race of life.
Abraham Lincoln

If a man does not keep pace with his companions, perhaps it is because he hears a different drummer. Let him step to the music which he hears, however measured or far away.
Henry Thoreau

At the time of the attack on the castle at Shimabara, Tazaki Geki was wearing very resplendent armor. Lord Katsushige was not pleased by this, and after that every time he saw something showy he would say, That's just like Geki's armor." In the light of this story, military armor and equipment that are showy can be seen as being weak and having no strength. By them one can see through the wearer's heart.
From the Hagakure

As long as people believe in absurdities, they will continue to commit atrocities.
Voltaire

Feeling deeply the difference between oneself and others, bearing ill will and falling out with people--these things come from a heart that lacks compassion. If one wraps up everything with a heart of compassion, there will be no coming into conflict with people.
From the Hagakure

Violence is the last refuge of the incompetent.
Issac Asimov

There is surely nothing other than the single purpose of the present moment. A man's whole life is a succession of moment after moment. If one fully understands the present moment, there will be nothing else to do, and nothing else to pursue. Live being true to the single purpose of the moment.
From the Hagakure

These are the levels in general; But there is one transcending level, and this is the most excellent of all. This person is aware of the endlessness of entering deeply into a certain Way and never thinks of himself as having finished. He truly knows his own insufficiencies and never in his whole life thinks that he has succeeded. He has no thoughts of pride but with self-abasement knows the Way to the end. It is said that Master Yagyu once remarked, I do not know the way to defeat others, but the way to defeat myself.
From the Hagakure

Men of high position, low position, deep wisdom and artfulness all feel that they are the ones who are working righteously, but when it comes to the point of throwing away one's life for his lord, all get weak in the knees. This is rather disgraceful. The fact that a useless person often becomes a matchless warrior at such times is because he has already given up his life and has become one with his master.
From the Hagakure

True patriotism hates injustice in its own land more than anywhere else.
Clarence Darrow

Patriotism is not short, frenzied outbursts of emotion, but the tranquil and steady dedication of a lifetime.
Adlai Stevenson

One of the great attractions of patriotism - it fulfills our worst wishes. In the person of our nation we are able, vicariously, to bully and cheat. Bully and cheat, what's more, with a feeling that we are profoundly virtuous.
Aldous Huxley

Speaking the Truth in times of universal deceit is a revolutionary act.
George Orwell

The more individuals capable of watching the world theater calmly and critically, the less danger of monumental mass stupidities in first of all, wars.
Hermann Hesse

My instinct as an individualist and artist has always warned me most urgently against this capacity of men for becoming drunk on collective suffering, collective pride, collective hatred, and collective honor. When this morbid exaltation becomes perceptible in a room, a hall, a village, a city, or a country, I grow cold and distrustful; a shudder comes over me, for already, while most of my fellow men are still weeping with rapture and enthusiasm, still cheering and venting protestations of brotherhood, I see blood flowing and cities going up in flames.
Hermann Hesse

There is one tactical principal which is not subject to change. It is to use the means at hand to inflict the maximum amount of wounds, death and destruction on the enemy in the minimum amount of time.
General George S. Patton

Nothing focuses the mind and cleanses the soul so well as facing your own execution in the morning.
Joseph Lumpkin

It is sweet and honorable to die for your country.
Horace

If ever there was a holy war, it was that which saved our liberties and gave us independence.
Thomas Jefferson

In the long run luck is given only to the efficient.
Helmuth von Moltke

The character of a soldier is high. They who stand forth the foremost in danger, for the community, have the respect of mankind. An officer is much more respected than any other man who has as little money. In a commercial country, money will always purchase respect. But you find, an officer, who has, properly speaking, no money, is everywhere well received and treated with attention. The character of a soldier always stands him in good stead.
Samuel Johnson

The concentration of troops can be done fast and easy, on paper.
Field Marshal Radomir Putnik

Victory in war does not depend entirely on numbers or courage; only skill and discipline will ensure it.
Flavius Vegetius

Wars may be fought with weapons, but they are won by men. It is the spirit of men who follow and of the man who leads that gains the victory.
General George S. Patton

What is our aim? Victory, victory at all costs, victory in spite of all terror; Victory how ever long and hard the road may be.
Sir Winston Churchill

In war there is no substitute for victory.
General Douglas MacArthur

There is one source, O Athenians, of all your defeats. It is that your citizens have ceased to be soldiers.
Demosthenes

Older men declare war. But it is the youth that must fight and die.
Herbert Hoover

Once we have a war there is only one thing to do. It must be won. For defeat brings worse things than any that can ever happen in war.
Ernest Hemingway

The sinews of war are not gold, but good soldiers; for gold alone will not procure good soldiers, but good soldiers will always procure gold.
Niccolò Machiavelli

I suppose every man is shocked when he hears how frequently soldiers are wishing for war. The wish is not always sincere; the greater part are content with sleep and lace, and counterfeit an ardor which they do not feel; but those who desire it most are neither prompted by malevolence nor patriotism; they neither pant for laurels, nor delight in blood; but long to be delivered from the tyranny of idleness, and restored to the dignity of active beings.   Samuel Johnson

The art of war is, in the last result, the art of keeping one's freedom of action
Xenophon

When bad men combine, the good must associate else they will fall one by one, an unpitied sacrifice in a contemptible struggle.
Edmund Burke

A soldier's time is passed in distress and danger, or in idleness and corruption.
Samuel Johnson

The Character of War

WHITE CLIFFS OF DOVER
World War II Song by Nat Burton and
Walter Kent, 1941

There'll be bluebirds over,
The white cliffs of Dover,
Tomorrow, just you wait and see.

There'll be love and laughter, And peace ever after,
Tomorrow when the world is free.

Courage, fearlessness, intelligence, cunning, bravery, integrity, discipline, humility, faithfulness, confidence, valor….all describe the indescribable.

The character of war is all of these yet more; all of these yet less. One single man or one infinite army, one hero or one despicable tyrant…alike yet completely disparate…imbued with the call to conquer, the mission to dominate.

- The Editors

There are three essentials to leadership: humility, clarity and courage.
Fuchan Yean

A leader is a dealer in hope.
Napoleon Bonaparte

I don't know what effect these men will have on the enemy, but, by God, they terrify me.
Duke of Wellington

Those in supreme power always suspect and hate their next heir.
Tacitus

Reason and judgment are the qualities of a leader.
Tacitus

Life is hard. Life is harder if you're stupid.
John Wayne

A little man often cast a long shadow.
G. M. Trevelyan

Action springs not from thought, but from a readiness for responsibility.
G. M. Trevelyan

Anger is a momentary madness, so control your passion or it will control you.
G. M. Trevelyan

It is not those who can inflict the most, but those that can suffer the most who will conquer.
Terence MacSwiney

More has been screwed up on the battlefield and misunderstood in the Pentagon because of a lack of understanding of the English language than any other single factor.
John W. Vessey, Jr.

Our strategy is one of preventing war by making it self-evident to our enemies that they're going to get their clocks cleaned if they start one.
John W. Vessey, Jr.

The courage of a soldier is heightened by his knowledge of his profession.
Flavius Renatus Vegetius

We find that the Romans owed the conquest of the world to no other cause than continual military training, exact observance of discipline in their camps, and unwearied cultivation of the other arts of war.
Flavius Renatus Vegetius

If God wanted us to be brave, why did he give us legs?
Marvin Kitman

There will be times when we think hard work and training are of no use. We will blame our lot in life on fate alone. In our despair we may say: Valor is of no service, chance rules all, and the bravest often fall by the hands of cowards.
Tacitus

I think with the Romans, that the general of today should be the soldier of tomorrow, if necessary.
Thomas Jefferson

To justify a fault is to argue for your own downfall.
Joseph Lumpkin

After I, a man of little rank, unexpectedly took control of the province, I have put forth great effort both day and night, at one time to gather together famous men of all kinds, listened to what they had to say, and have continued in such a way up to this time.
Asakura Toshikage

One matures into leadership. It overtakes him as he learns and expands. Those who seek leadership usually do so prematurely. When people see those things in you they desire in themselves, they will follow. Only then are you a leader.
Joseph Lumpkin

It is the business of a general to be quiet and thus ensure secrecy; and to be upright and just, and thus maintain order.
Sun Tzu

The supreme quality for leadership is unquestionably integrity. Without it, no real success is possible, no matter whether it is on a section gang, a football field, in an army, or in an office.
General Dwight D. Eisenhower

Those who desire to govern their states should first put their families in order. And those who desire to put their families in order would first discipline themselves.
Confucius

The consummate leader cultivates the moral law, and strictly adheres to method and discipline; thus it is in his power to control success.
Sun Tzu

Character is like a tree and reputation like a shadow. The shadow is what we think of it; the tree is the real thing.
Abraham Lincoln

The quality of decision is like the well-timed swoop of a falcon, which enables it to strike and destroy its victim. Therefore the good fighter will be terrible in his onset, and prompt in his decision. Energy may be likened to the bending of a crossbow; decision, to the releasing of a trigger.
Sun Tzu

Training is the education of instinct.
Anonymous

In this uncertain world, ours should be the path of discipline.
Hiba Yoshimasa

One should exert himself in martial arts absolutely. There are no weak soldiers under a strong general.
Takeda Nobushige

There is no deadlier weapon than the will! The sharpest sword is not equal to it. There is no robber so dangerous as nature. Yet, it is not nature that does the damage: it is man's own will!
Chuang Tzu

The fatal flaw of one promoted to a position of authority is to forget from where he came. If he remembers his previous low position he will see that every fool has a chance to advance. He will understand it is by grace and chance that he is there. If he understands there are many as good as he, the position will not seem so high and he will know all men are replaceable. This will keep him humble. In his humility he will treat others well and they will follow him willingly.
Joseph Lumpkin

Like everyone else, you want to learn the way to win, but never to accept the way to lose. To accept defeat - to learn to die - is to be liberated from it. Once you accept, you are free to flow and to harmonize.
Bruce Lee

I will stand off the forces of the entire county here, and die a glorious death.
Torii Mototada

Not being tense, but ready. Not thinking yet not dreaming, not being set, but flexible - it is being wholly and quietly alive, aware and alert, ready for whatever may come.
Bruce Lee

A person's character and depth of mind is seen by his behavior. Thus, one should understand that even the walls and fences have eyes... one should not take a single step in vain, or speak a word in a way that others may speak of him as shallow.
Shiba Yoshimasa

The wise adapt themselves to circumstances, as water molds itself to the pitcher.
Chinese proverb

Good leaders must first become good servants. Prosperity is a great teacher; adversity a greater.
William Hazlitt

Courage is doing what you're afraid to do. There can be no courage unless you're scared.
Eddie Rickenbacker

I cannot hear what you say for the thunder of what you are.
Zulu Proverb

Beware of the leader who bangs the drums of war to whip the citizenry into a patriotic fever. For patriotism is indeed a double edged sword. It both emboldens the blood, just as it narrows the mind. When the drums of war have reached a fever pitch, and the blood boils with hate and the mind is closed, the leader will have no need in seizing the rights of the citizenry. Rather the citizenry, infused with fear and blinded by patriotism, will offer up all of their rights unto the leader and do it gladly so. How do I know? I know for this is what I have done. And I am Caesar.
William Shakespeare

Duty is what one expects from others.
Oscar Wilde

Only those who dare to fail greatly can ever achieve greatly.
Robert F. Kennedy

As for courage and will - we cannot measure how much of each lies within us, we can only trust there will be sufficient to carry through trials which may lie ahead.
Andre Norton

Disobedience, the rarest and most courageous of the virtues, is seldom distinguished from neglect, the laziest and commonest of the vices.
George Bernard Shaw

You may be deceived if you trust too much, but you will live in torment if you don't trust enough.
Frank H. Crane

Courage is grace under pressure.
Ernest Hemingway

If you can talk brilliantly about a problem, it can create the consoling illusion that it has been mastered.
Stanley Kubrick

The world breaks everyone, and afterward, some are stronger at the broken places.
Ernest Hemingway

It is hard to fight an enemy who has outposts in your head.
Sally Kempton

Life is what happens to us while we're making other plans.
Thomas LaMance

Leaders get out in front and stay there by raising the standards by which they judge themselves - and by which they are willing to be judged.
Fredrick Smith

A successful man is one who can lay a firm foundation with the bricks that others throw at him.
Sidney Greenberg

Good leaders must first become good servants.
Robert Greenleaf

One does not discover new lands without consenting to lose sight of the shore for a very long time.
André Gide

If you want to piss with the big dogs, you'd better learn to lift your leg first; otherwise you just might get pissed on.
E.M. Glenn

It's there within us all. It costs nothing, takes almost no time and is powerful beyond measure. Unleash the power of praise and reap the rewards.
Susan Mitchell

Better to remain silent and be thought a fool than to speak out and remove all doubt.
Abraham Lincoln

I'm a slow walker, but I never walk back.
Abraham Lincoln

It's the process of striving that makes us grow - not necessarily the result. Running against the wind makes us better than running with it.
Herb Elliott

Courage is the finest of human qualities because it is the quality which guarantees all others.
Sir Winston Churchill

Courage - a perfect sensibility of the measure of danger, and a mental willingness to endure it.
William Tecumseh Sherman

Courage, hard work, self-mastery, and intelligent effort are all essential to successful life.
Theodore Roosevelt

What we need in this country today is more courage and more belief in the things that we have.
Thomas Watson

A lot of people do not muster the courage to live their dreams because they are afraid to die.
Les Brown

It is curious that physical courage should be so common in the world and moral courage so rare.
Mark Twain

Hope, like faith, is nothing if it is not courageous; it is nothing if it is not ridiculous.
Thornton Wilder

Courage consists not in hazarding without fear, but being resolutely minded in a just cause.
Plutarch

It is stupidity rather than courage to refuse to recognize danger when it is close upon you.
Sir Arthur Conan Doyle

Support the strong, give courage to the timid, remind the indifferent, and warn the opposed.
Whitney M. Young

It's not the maker of the sword, but the courage and skill of the swordsman that wins the day.
Robert M. Irwin

The only kind of courage that matters is the kind that gets you from one minute to the next.
Mignon McLaughlin

You cannot build character and courage by taking away men's initiative and independence.
William J. H. Boetcker

It takes as much courage to have tried and failed as it does to have tried and succeeded.
Anne Lindbergh

The courage we desire and prize is not the courage to die decently, but to live manfully.
Thomas Carlyle

Hope awakens courage. He who can implant courage in the human soul is the best physician.
Karl Ludwig von Knebel

Courage is doing without witnesses that which we would be capable of doing before everyone.
Duke de La Rochefoucauld

To call war the soil of courage and virtue is like calling debauchery the soil of love.
George Santayana

A good man will certainly also possess courage; but a brave man is not necessarily good.
Confucius

Never ask the Gods for life set free from grief, but ask for courage that endureth long.
Menander

Courage consists not in blindly overlooking danger, but in seeing it, and conquering it.
Jean Paul Richter

It takes vision and courage to create - it takes faith and courage to prove.
Owen D. Young

The greatest test of courage on earth is to bear defeat without losing heart.
Robert Ingersoll

He who is not courageous enough to take risks will accomplish nothing in life.
Muhammad Ali

Good ideas and innovations must be driven into existence by courage and patience.
Admiral Hyman G. Rickover

We must have courage to bet on our ideas, to take the calculated risk, and to act.
Maxwell Maltz

You cannot discover new oceans unless you have the courage to lose sight of the shore.
Unknown

All you need is the plan, the road map, and the courage to press on to your destination.
Earl Nightingale

To see what is right, and not do it, is want of courage, or of principle.
Confucius

We must constantly build dikes of courage to hold back the flood of fear.
Martin Luther King Jr.

Success is never final and failure never fatal. It's courage that counts.
George F. Tilton

True courage is a result of reasoning. A brave mind is always impregnable.
Jeremy Collier

Courage is the strong desire to live taking the form of a readiness to die.
G.K. Chesterton

Courage is resistance to fear, mastery of fear - not absence of fear.
Mark Twain

Some temptations are so great it takes great courage to yield to them.
Oscar Wilde

A great part of courage is the courage of having done the thing before.
Ralph Waldo Emerson

The secret of Happiness is Freedom, and the secret of Freedom, Courage.
Thucydides

Courage conquers all things; it even gives strength to the body.
Ovid

Efforts and courage are not enough without purpose and direction.
John F. Kennedy

Courage is the capacity to confirm what can be imagined.
Leo Calvin Rosten

Confidence is directness and courage in meeting the facts of life.
John Dewey

Failure is unimportant. It takes courage to make a fool of oneself.
Charlie Chaplin

If we survive danger it steels our courage more than anything else.
Reinhold Niebuhr

Few persons have courage enough to appear as good as they really are.
Augustus Hare

Most men have more courage than even they themselves think they have.
Lord Brook Fulke Greville

Courage is like love - it must have hope to nourish it.
Napoleon Bonaparte

Courage is being scared to death - and saddling up anyway.
John Wayne

It takes courage to grow up and become who you really are.
e.e. cummings

Courage is the ladder on which all the other virtues mount.
Clare Boothe Luce

Nothing gives a fearful man more courage than another's fear.
Umberto Eco

Courage is simply the willingness to be afraid and act anyway.
Dr. Robert Anthony

Until the day of his death, no man can be sure of his courage.
Jean Anouilh

True courage is like a kite; a contrary wind raises it higher.
John Petit-Senn

One man with courage makes a majority.
Andrew Jackson

Without courage, wisdom bears no fruit.
Baltasar Gracian

Courage is fire, and bullying is smoke.
Benjamin Disraeli

Courage without conscience is a wild beast.
Robert Ingersoll

Courage is fear holding on a minute longer.
General George S. Patton

Courage leads to heaven; fear leads to death.
Seneca

Fortune can take away riches, but not courage.
Seneca

It requires more courage to suffer than to die.
Napoleon Bonaparte

It is in great dangers that we see great courage.
Jean Francois Regnard

Life shrinks or expands in proportion to one's courage.
Anais Nin

These are the times that try men's souls. The summer soldier and the sunshine patriot will, in this crisis, shrink from the service of his country; but he that stands it now, deserves the love and thanks of man and woman. Tyranny, like hell, is not easily conquered; yet we have this consolation with us, that the harder the conflict, the more glorious the triumph. What we may obtain too cheap, we esteem too lightly.
Thomas Paine

Courage follows action.
Mack R. Douglas

Have the courage to be wise.
Horatius

Without justice, courage is weak.
Benjamin Franklin

You can't test courage cautiously.
Annie Dillard

Necessity does the work of courage.
George Eliot

Courage in danger is half the battle.
Titus Maccius Plautus

Every man who expresses an honest thought is a soldier in the army of intellectual liberty.
Robert Ingersoll

To be a successful soldier you must know history. What you must know is how man reacts. Weapons change but the man who uses them changes not at all. To win battles you do not beat weapons - you beat the soul of man of the enemy man.
General George S. Patton

Leadership is solving problems. The day soldiers stop bringing you their problems is the day you have stopped leading them. They have either lost confidence that you can help or concluded you do not care. Either case is a failure of leadership.
Karl Popper

I have seen soldiers panic at the first sight of battle, and a wounded squire pulling arrows out from his wound to fight and save his dying horse. Nobility is not a birthright but is defined by one's action.
Robin Hood, Prince of Thieves

Valor, glory, firmness, skill, generosity, steadiness in battle and ability to rule - these constitute the duty of a soldier. They flow from his own nature.
Bhagavad Gita

The most vital quality a soldier can possess is self-confidence, utter, complete and bumptious.
General George S. Patton

The dignity of man is vindicated as much by the thinker and poet as by the statesman and soldier.
James Bryant Conant

No matter whether a person belongs to the upper or lower ranks, if he has not put his life on the line at least once he has cause for shame.
Nabeshima Naoshige

Being affected by the avarice for office and rank, or wanting to become a daimyo and being eager for such things ... will not one then begin to value his life? And how can a man commit acts of martial valor if he values his life? A man who has been born into the house of a warrior and yet places no loyalty in his heart and thinks only of the fortune of his position will be flattering on the surface and construct schemes in his heart, will forsake righteousness and not reflect on his shame, and will stain the warrior's name of his household to later generations. This is truly regrettable.
Torii Mototada

Life is like unto a long journey with a heavy burden. Let thy step be slow and steady, that thou stumble not. Persuade thyself that imperfection and inconvenience are the natural lot of mortals, and there will be no room for discontent, neither for despair. When ambitious desires arise in thy heart, recall the days of extremity thou has passed through. Forbearance is the root of quietness and assurance forever. Look upon the wrath of the enemy. If thou knowest only what it is to conquer, and knowest not what it is to be defeated, woe unto thee; it will fare ill with thee. Find fault with thyself rather than with others.
Tokugawa

In strategy your spiritual bearing must not be any different from normal. Both in fighting and in everyday life you should be determined though calm. Meet the situation without tenseness yet not recklessly, your spirit settled yet unbiased. If the enemy thinks of the mountains, attack like the sea; and if he thinks of the sea, attack like the mountains.
Miyamoto Musashi

The combining of these three virtues may seem unobtainable to the ordinary person, but it is easy. Intelligence is nothing more than discussing things with others. Wisdom comes from this. Humanity is something done for the sake of others, simply comparing oneself with them and putting them in the fore. Courage is gritting one's teeth ; it is simply doing that and pushing ahead, paying no attention to the circumstances. Anything that seems above these three is not necessary to be known.
From the Hagakure

When one is attending to matters, there is one thing that comes forth from his heart. That is, in terms of one's lord, loyalty; in terms of one's parents, filial piety; in martial affairs, bravery; and apart from that, something that can be used by all the world.
From the Hagakure

The essential American character is hard, isolated, stoic, and a killer.
D.H. Lawrence

In a civil war, a general must know exactly when to move over to the other side.
Henry Reed

Patriotism is the virtue of the vicious.
Oscar Wilde

A really great people, proud and high-spirited, would face all the disasters of war rather than purchase that base prosperity which is bought at the price of national honor.
Theodore Roosevelt

No man can sit down and withhold his hands from the warfare against wrong and get peace from his acquiescence.
Woodrow Wilson

No person was ever honored for what he received. Honor has been the reward for what he gave.
Calvin Coolidge

Never in the field of human conflict was so much owed by so many to so few.
Sir Winston Churchill

Let us solemnly remember the sacrifices of all those who fought so valiantly, on the seas, in the air, and on foreign shores, to preserve our heritage of freedom, and let us re-consecrate ourselves to the task of promoting an enduring peace so that their efforts shall not have been in vain.
General Dwight D. Eisenhower

Cowards die many times before their deaths; the valiant never taste of death but once.
William Shakespeare

To lead uninstructed people to war is to throw them away.
Confucius

The secret of all victory lies in the organization of the non-obvious.
Marcus Aurelius

I believe that military service in the Armed Forces of the United States is a profound form of service to all humankind. You stand engaged in an effort to keep America safe at home, to protect our allies and interests abroad, to keep the seas and the skies free of threat. Just as America stands as an example to the world of the inestimable benefits of freedom and democracy, so too an America with the capacity to project her power for the purpose of protecting and expanding freedom and democracy abroad benefits the suffering people of the world.
Ronald Reagan

Time is a sort of river of passing events, and strong is its current; no sooner is a thing brought to sight than it is swept by and another takes its place, and this too will be swept away.
Marcus Aurelius

Leadership is the art of getting someone else to do something you want done because he wants to do it.
General Dwight D. Eisenhower

Nearly all men can stand adversity, but if you want to test a man's character, give him power.
Abraham Lincoln

LOVE AND WAR

DANNY BOY
Lyrics by Fred Weatherly, 1911

Oh, Danny boy, the pipes, the pipes are calling
From glen to glen, and down the mountain side.
The summer's gone, and all the roses falling,
It's you, it's you must go and I must bide.
But come ye back when summer's in the meadow,
Or when the valley's hushed and white with snow,
It's I'll be here in sunshine or in shadow,
Oh, Danny boy, O Danny boy, I love you so!

But when ye come, and all the flowers are dying,
If I am dead, as dead I well may be.
Ye'll come and fine the place where I am lying,
And kneel and say an Ave' there for me.
And I shall hear, though soft you tread above me,
And all my grave will warmer, sweeter be,
For you will bend and tell me that you love me,
And I shall sleep in peace until you come to me.

The rush of submission blends with the power of victory. Yet in the final analysis there are winners and there are losers and the survivors are forever damaged and scarred. Love and war...war and love; inextricably joined; painfully entwined, one with the other, for all eternity.

The Editors

Love is like war: easy to begin but very hard to stop.
H. L. Mencken

Love does not begin and end the way we seem to think it does.
Love is a battle, love is a war; love is a growing up.
James A. Baldwin

The Wedding March always reminds me of the music played
when soldiers go into battle.
Heinrich Heine

Nothing is miserable unless you think it is so.
Boethius

War is like love; it always finds a way.
Bertold Brecht

For in all adversity of fortune the worst sort of misery is to have
been happy.
Boethius

Love is an ocean of emotions entirely surrounded by expenses.
Lord Thomas Dewar

The art of love is largely the art of persistence.
Albert Ellis

Fortune and love favor the brave.
Ovid

Love conquers all.
Virgil

Love is a canvas furnished by nature and embroidered by imagination.
Voltaire

Love is a net that catches hearts like a fish.
Muhammad Ali

Love is the only force capable of transforming an enemy into friend.
Martin Luther King, Jr.

Man must evolve for all human conflict a method which rejects revenge, aggression and retaliation. The foundation of such a method is love.
Martin Luther King, Jr.

All brave men love; for he only is brave who has affections to fight for, whether in the daily battle of life, or in physical contests.
Nathaniel Hawthorne

All married couples should learn the art of battle as they should learn the art of making love. Good battle is objective and honest - never vicious or cruel. Good battle is healthy and constructive, and brings to a marriage the principles of equal partnership.
Ann Landers

A woman watches her body uneasily, as though it were an unreliable ally in the battle for love.
Leonard Cohen

All the passions make us commit faults; love makes us commit the most ridiculous ones.
Duke de La Rochefoucauld

And yet a little tumult, now and then, is an agreeable quickener of sensation; such as a revolution, a battle, or an adventure of any lively description.
Lord Byron

Marriage is an adventure, like going to war.
Gilbert K. Chesterton

Men like war: they do not hold much sway over birth, so they make up for it with death. Unlike women, men menstruate by shedding other people's blood.
Lucy Ellman

Power is my mistress. I have worked too hard at her conquest to allow anyone to take her away from me.
Napoleon Bonaparte

War has always been the grand sagacity of every spirit which has grown too inward and too profound; its curative power lies even in the wounds one receives.
Friedrich Nietzsche

Love of country is like love of woman--he loves her best who seeks to bestow on her the highest good.
Felix Adler

You say that love is nonsense....I tell you it is no such thing. For weeks and months it is a steady physical pain, an ache about the heart, never leaving one, by night or by day; a long strain on one's nerves like toothache or rheumatism, not intolerable at any one instant, but exhausting by its steady drain on the strength.
Henry Brooks Adams

In dreams and in love there are no impossibilities.
Janos Arany

And in the end, the love you take is equal to the love you make.
The Beatles

The only way of knowing a person is to love them without hope.
Walter Benjamin

Love is very patient, Love is very kind, Love is never envious. Or vaunted up with pride.
Holy Bible, Book of 1st Corinthians

Love is always either increasing or decreasing.
Andreas Capellanus

We always deceive ourselves twice about the people we love - first to their advantage, then to their disadvantage.
Albert Camus

Give me more love or more disdain; The torrid or the frozen zone; Bring equal ease unto my pain; The temperate affords me none.
Thomas Carew

Of all the pain, the greatest pain is to love, but to love in vain.
Abraham Cowley

Love is a power too strong to be overcome by anything but flight.
Miguel de Cervantes

There are people who would have never fallen in love if they never heard of love.
Duke de La Rochefoucauld

Love never dies of starvation, but often of indigestion.
Ninon de Lenclos

Love does not consist in gazing at each other but in looking together in the same direction.
Antoine de Saint-Exupery

Love is an irresistible desire to be irresistibly desired.
Robert Frost

It is the special quality of love not to be able to remain stationary, to be obliged to increase under pain of diminishing.
Andre' Gide

Love is a perky elf dancing a merry little jig and then suddenly he turns on you with a miniature machine-gun.
Matt Groening

Never judge someone by who he's in love with; judge him by his friends. People fall in love with the most appalling people.
Cynthia Heimel

Love begets love, love knows no rules, this is the same for all.
Virgil

Hatred paralyzes life; love releases it. Hatred confuses life; love
harmonizes it. Hatred darkens life; love illuminates it.
Martin Luther King, Jr.

Love is the delusion that one woman differs from another.
H. L. Mencken

Love is the triumph of imagination over intelligence.
H. L. Mencken

To be in love is merely to be in a state of perpetual anesthesia -
to mistake an ordinary young woman for a goddess.
H. L. Mencken

Alas! how light a cause may move
Dissension between hearts that love!
Hearts that the world in vain had tried,
And sorrow but more closely tied;
That stood the storm when waves were rough,
Yet in a sunny hour fall off.
Thomas Moore

Love is much like a wild rose, beautiful and calm, but willing to
draw blood in its defense.
Mark A. Overby

Love and dignity cannot share the same abode.
Ovid

Love is like quicksilver in the hand. Leave the fingers open and
it stays. Clutch it, and it darts away.
Dorothy Parker

Love is a reciprocal torture.
Marcel Proust

Love is like the moon; when it does not increase it decreases.
Segur

Men have died from time to time, and the worms have eaten
them, but not for love.
William Shakespeare

First love is only a little foolishness and a lot of curiosity.
George Bernard Shaw

All is fair in love and war.
Francis Edward Smedley

The joy of late love is like green firewood when set aflame, for
the longer the wait in lighting, the greater heat it yields and the
longer its force lasts.
Chrétien de Troyes

It has ever been since time began,
And ever will be, till time lose breath,
That love is a mood - no more - to man,
And love to a woman is life or death.
Ella Wheeler Wilcox

Yet each man kills the thing he loves,
By each let this be heard,
Some do it with a bitter look
Some with a flattering word,
The coward does it with a kiss,
The brave man with a sword!
Oscar Wilde

Warriors on War

THE BATTLE CRY OF FREEDOM
Civil War Song, George F. Root, 1861

Yes, we'll rally round the flag, boys,
We'll rally once again,
Shouting the battle-cry of Freedom;
We will rally from the hillside,
We will gather from the plain,
Shouting the battle-cry of Freedom.

Onward in infinite legions they march, wave after wave...the centurians, the magyars, the mongols, the samurai, the chevaliers, the hwa rang, the green berets...all breathing fire, all eyes forward; to glory or to death, into the teeth of battle. What do they think and feel, these human machines of war – do we detect fear? Regret? Joy? Rage? The overpowering rush of adrenalin, and then it begins.

- The Editors

Cry 'Havoc', and let slip the dogs of war.
William Shakespeare

Minds are like parachutes. They function only when they are open!
Lord Dewar

One 'Oh shit' wipes out 30 'Atta boys'!
U.S. Marine Corps

If you want a decision, go to the point of danger.
General James M. Gavin

When we jumped into Sicily, the units became separated, and I couldn't find anyone. Eventually I stumbled across two colonels, a major, three captains, two lieutenants, and one rifleman, and we secured the bridge. Never in the history of war have so few been led by so many.
General James M. Gavin

Onward we stagger, and if the tanks come, may God help the tanks.
Colonel William O. Darby

The essence of a general's job is to assist in developing a clear sense of purpose to keep the junk from getting in the way of important things.
Lt. General Walter F. Ulmer

You can kill ten of my men for every one I kill of yours, but even at those odds, you will lose and I will win.
Ho Chi Minh

Intuitive decision-making and mastering this profession are one in the same.
Lt. General Van Riper

REVEILLE, n. A signal to sleeping soldiers to dream of battlefields no more, but get up and have their blue noses counted.
Ambrose Bierce

DRAGOON, n. A soldier who combines dash and steadiness in so equal measure that he makes his advances on foot and his retreats on horseback.
Ambrose Bierce

The purpose of studying the new sciences is simple... We want to learn to understand war through the most powerful means available... to encompass the ideas contained in quantum mechanics, nonlinear systems, and chaos and complexity theories.
Lt. General Van Riper

This is an era of violent peace.
Admiral James D. Watkins

An extraordinary affair. I gave them their orders and they wanted to stay and discuss them.
Duke of Wellington

War is fear cloaked in courage.
General William C. Westmoreland

War can only be abolished through war, and in order to get rid of the gun it is necessary to take up the gun.
Mao Tse Tung

It is a fact that under equal conditions, large-scale battles and whole wars are won by troops which have a strong will for victory, clear goals before them, high moral standards, and devotion to the banner under which they go into battle.
Marshal Georgi Zhukov

The nature of encounter operations require of the commanders limitless initiative and constant readiness to take the responsibility for military actions.
Marshal Georgi Zhukov

It is better to die on your feet than to live on your knees.
Emiliano Zapata

An army without culture is a dull-witted army, and a dull-witted army cannot defeat the enemy.
Mao Tse Tung

A man who has been shot at is a new realist, and what do you say to a realist when the war is a war of ideals?
Michael Shaara

It's the idea that we all have value, you and me, and we're worth something more than the dirt. I never saw dirt I'd die for, but I'm not asking you to come join us and fight for dirt. What we're all fighting for, in the end, is each other.
Michael Shaara

They would fight again, and when they came he would be behind another stone wall waiting for them, and he would stay there until he died or until it ended, and he was looking forward to it with an incredible eagerness, as you wait for the great music to begin after the silence.
Michael Shaara

Humility must always be the portion of any man who receives acclaim earned in the blood of his fellows and the sacrifice of his friends.
General Dwight D. Eisenhower

Comrades, you have lost a good captain to make a bad general.
Saturninus

Leadership is a combination of strategy and character. If you must be without one, be without the strategy.
General H. Norman Schwarzkopf

The transition from the defensive to the offensive is one of the most delicate operations of war.
Napoleon Bonaparte

When an archer is shooting for nothing he has all of his skill. If he shoots for a brass buckle he is already nervous. If he shoots for a prize of gold he goes blind... his skill has not changed, but the prize divides him.
Chuang Tzu

As far as Saddam Hussein being a great military strategist, he is neither a strategist, nor is he schooled in the operational arts, nor is he a tactician, nor is he a general, nor is he a soldier. Other than that, he's a great military man. I want you to know that.
General H. Norman Schwarzkopf

Do what is right, not what you think the high headquarters wants or what you think will make you look good.
General H. Norman Schwarzkopf

It doesn't take a hero to order men into battle. It takes a hero to be one of those men who goes into battle.
General H. Norman Schwarzkopf

The day we executed the air campaign, I said, we gotcha!
General H. Norman Schwarzkopf

When placed in command - take charge.
General H. Norman Schwarzkopf

A good plan violently executed now is better than a perfect plan executed next week.
General George S. Patton

A piece of spaghetti or a military unit can only be led from the front end.
General George S. Patton

A pint of sweat saves a gallon of blood.
General George S. Patton

All very successful commanders are prima donnas and must be so treated.
General George S. Patton

Always do everything you ask of those you command.
General George S. Patton

Battle is an orgy of disorder.
General George S. Patton

It is foolish and wrong to mourn the men who died. Rather we should thank God that such men lived.
General George S. Patton

Battle is the most magnificent competition in which a human being can indulge. It brings out all that is best; it removes all that is base. All men are afraid in battle. The coward is the one who lets his fear overcome his sense of duty. Duty is the essence of manhood.
General George S. Patton

Better to fight for something than live for nothing.
General George S. Patton

If we take the generally accepted definition of bravery as a quality which knows no fear, I have never seen a brave man. All men are frightened. The more intelligent they are, the more they are frightened.
General George S. Patton

The object of war is not to die for your country, but to make the other bastard die for his.
General George S. Patton

Nobody ever defended anything successfully, there is only attack and attack and attack some more.
General George S. Patton

Take calculated risks. That is quite different from being rash.
General George S. Patton

The time to take counsel of your fears is before you make an important battle decision. That's the time to listen to every fear you can imagine! When you have collected all the facts and fears and made your decision, turn off all your fears and go ahead!
General George S. Patton

There is only one sort of discipline, perfect discipline.
General George S. Patton

Untutored courage is useless in the face of educated bullets.
General George S. Patton

War is an art and as such is not susceptible of explanation by fixed formula.
General George S. Patton

We herd sheep, we drive cattle, we lead people. Lead me, follow me, or get out of my way.
General George S. Patton

A general is just as good or just as bad as the troops under his command make him.
General Douglas MacArthur

And like the old soldier in that ballad, I now close my military career and just fade away, an old soldier who tried to do his duty as God gave him the sight to see that duty.
General Douglas MacArthur

In my dreams I hear again the crash of guns, the rattle of musketry, the strange, mournful mutter of the battlefield.
General Douglas MacArthur

In war, you win or lose, live or die - and the difference is just an eyelash.
General Douglas MacArthur

It is fatal to enter any war without the will to win it.
General Douglas MacArthur

Never give an order that can't be obeyed.
General Douglas MacArthur

The best luck of all is the luck you make for yourself.
General Douglas MacArthur

The outfit soon took on color, dash and a unique flavor which is the essence of that elusive and deathless thing called soldiering.
General Douglas MacArthur

We are not retreating - we are advancing in another direction.
General Douglas MacArthur

Whether in chains or in laurels, liberty knows nothing but victories.
General Douglas MacArthur

Last, but by no means least, courage - moral courage, the courage of one's convictions, the courage to see things through. The world is in a constant conspiracy against the brave. It's the age-old struggle; the roar of the crowd on one side and the voice of your conscience on the other.
General Douglas MacArthur

A true man of honor feels humbled himself when he cannot help humbling others.
General Robert E. Lee

Duty is the sublimest word in the language. You can never do more than your duty. You should never wish to do less.
General Robert E. Lee

It is well that war is so terrible. We should grow too fond of it.
General Robert E. Lee

Let the tent be struck.
General Robert E. Lee

Never do a wrong thing to make a friend or to keep one.
General Robert E. Lee

The education of a man is never completed until he dies.
General Robert E. Lee

We failed, but in the good providence of God apparent failure often proves a blessing.
General Robert E. Lee

We have fought this fight as long, and as well as we know how. We have been defeated. For us as a Christian people, there is now but one course to pursue. We must accept the situation.
General Robert E. Lee

What a cruel thing is war: to separate and destroy families and friends, and mar the purest joys and happiness God has granted us in this world; to fill our hearts with hatred instead of love for our neighbors, and to devastate the fair face of this beautiful world.
General Robert E. Lee

Bravery is the capacity to perform properly even when scared half to death.
General Omar N. Bradley

I am convinced that the best service a retired general can perform is to turn in his tongue along with his suit and to mothball his opinions.
General Omar N. Bradley

If we continue to develop our technology without wisdom or prudence, our servant may prove to be our executioner.
General Omar N. Bradley

Leadership in the democratic army means firmness, not harshness; understanding, not weakness; justice, not license; humaneness, not intolerance; generosity, not selfishness; pride, not egotism.
General Omar N. Bradley

Leadership is intangible, and therefore no weapon ever designed can replace it.
General Omar N. Bradley

Ours is a world of nuclear giants and ethical infants. We know more about war than we know about peace, more about killing than we know about living.
General Omar N. Bradley

The way to win an atomic war is to make certain it never starts.
General Omar N. Bradley

Wars can be prevented just as surely as they can be provoked, and we who fail to prevent them, must share the guilt for the dead.
General Omar N. Bradley

We need to learn to set our course by the stars, not by the light of every passing ship.
General Omar N. Bradley

Do not needlessly endanger your lives until I give you the signal.
General Dwight D. Eisenhower

If a problem cannot be solved, enlarge it.
General Dwight D. Eisenhower

It is far more important to be able to hit the target than it is to haggle over who makes a weapon or who pulls a trigger.
General Dwight D. Eisenhower

Pessimism never won any battle.
General Dwight D. Eisenhower

There is no victory at bargain basement prices.
General Dwight D. Eisenhower

What counts is not necessarily the size of the dog in the fight - it's the size of the fight in the dog.
General Dwight D. Eisenhower

The art of war is simple enough. Find out where your enemy is. Get at him as soon as you can. Strike him as hard as you can, and keep moving.
General Ulysses S. Grant

To have good soldiers, a nation must always be at war.
Napoleon Bonaparte

The essence of war is violence. Moderation in war is imbecility.
British Sea Lord John Fisher

He who stays on the defensive does not make war, he endures it.
Field Marshal Colmar Baron von der Goltz

If men make war in slavish obedience to rules, they will fail.
General Ulysses S. Grant

War is the mother of everything.
Heraclitus

In war, only the simple succeed.
Field Marshal Paul Von Hindenburg

Si vis pacem, para bellum
(If you want peace, prepare for war!)
Flavius Vegetius Renatus

War is cruelty. There's no use trying to reform it, the crueler it is
the sooner it will be over.
William Tecumseh Sherman

The true test of a leader is whether his followers will adhere to
his cause from their own volition, enduring the most arduous
hardships without being forced to do so, and remaining
steadfast in the moments of greatest peril.
Xenophon

"What a piece of work is man…in action how like an angel!"
Well, boy, if he's an angel, he's sure a killer angel.
Michael Shaara

Knowing is not enough; we must apply. Willing is not enough;
we must do.
Bruce Lee

There is no standard in total combat, and expression must be
free. This liberating truth is a reality only in so far as it is
experienced and lived by the individual himself; it is a truth that
transcends styles or disciplines.
Bruce Lee

Humble words and increased preparations are signs that the enemy is about to advance. Violent language and driving forward as if to the attack are signs that he will retreat. When the light chariots come out first and take up a position on the wings, it is a sign that the enemy is forming for battle. Peace proposals unaccompanied by a sworn covenant indicate a plot. When there is much running about and the soldiers fall into rank, it means that the critical moment has come. When some are seen advancing and some retreating, it is a lure.
Sun Tzu

The general, unable to control his irritation, will launch his men to the assault like swarming ants, with the result that one-third of his men are slain, while the town still remains untaken. Such are the disastrous effects of a siege.
Sun Tzu

If we wish to fight, the enemy can be forced to an engagement even though he be sheltered behind a high rampart and a deep ditch. All we need do is attack some other place that he will be obliged to relieve.
Sun Tzu

Hold out baits to entice the enemy. Feign disorder, and crush him.
Sun Tzu

You intentionally leave an opening in your defenses or leave some limb vulnerable to attack in order to draw your opponent into making a predictable attack which you can in turn counter. Like baiting a fish hook.
Bruce Lee

Though a warrior may be called a dog or a beast, what is basic to his nature is to win.
Asakura Norikage

You can prevent your opponent from defeating you through defense, but you cannot defeat him without taking the offensive.
Sun Tzu

Show him there is a road to safety, and so create in his mind the idea that there is an alternative to death. Then, Strike!
Tu Mu quoted in the Art of War by Sun Tzu.

You can be sure of succeeding in your attacks if you only attack places, which are left open and not defended. You can ensure the safety of your defense if you only hold positions that cannot be attacked.
Sun Tzu

To fight and conquer in one hundred battles is not the highest skill. To subdue the enemy with no fight at all, that's the highest skill.
Sun Tzu

The non-action of the wise man is not inaction...The sage is quiet because he is not moved... The heart of the wise man is tranquil... emptiness, stillness, tranquility, tastelessness, silence, and non-action are the root of all things.
Chuang Tzu

The good fighters of old first put themselves beyond the possibility of defeat, and then waited for an opportunity of defeating the enemy.
Sun Tzu

Success in warfare is gained by carefully accommodating ourselves to the enemy's purpose.
Sun Tzu

There are many ways to describe a warrior. The descriptions may vary widely, so it may be easier to start with what a warrior is not. The warrior is never a brute. He is never reactionary. He is not arrogant or pompous. The warrior should be a balanced and educated person. Most of all, the warrior should have the strength of a strong man, the heart of a compassionate woman, and the open mind of a little child.
Joseph Lumpkin

A good soldier is not violent. A good fighter does not get angry. A good winner is never vengeful. A good employer is humble. This is the virtue of not striving. It is known as the ability to deal with people.
Lao Tzu

The path of the Warrior is lifelong, and mastery is often simply staying on the path.
Richard Strozzi Heckler

The Way of a Warrior is based on humanity, love, and sincerity; the heart of martial valor is true bravery, wisdom, love, and friendship. Emphasis on the physical aspects of warriorship is futile, for the power of the body is always limited.
Ueshiba Morihei

I'm not a good shot, but I shoot often.
Theodore Roosevelt

When I came back to Dublin I was court martialed in my absence and sentenced to death in my absence, so I said they could shoot me in my absence.
Brendan Behan

If you know the enemy and know yourself, you need not fear the result of a hundred battles. If you know yourself but not the enemy, for every victory gained you will also suffer a defeat. If you know neither the enemy nor yourself, you will succumb in every battle.
Sun Tzu

You smell that? Do you smell that? Napalm, son. Nothing else in the world smells like that. I love the smell of napalm in the morning. You know, one time we had a hill bombed, for twelve hours. When it was all over I walked up. We didn't find one of 'em, not one stinkin' dink body. The smell, you know that gasoline smell, the whole hill. Smelled like... victory.
Apocalypse Now

If you are going to deal in death, you should be willing to see the truth of it, not some glorious lie. If I have a battle with another sword player, it is between the two of us, our business, our truth. But if you run a planet and you get pissed off at somebody the next orbit over, you each might send a million soldiers to recycling plants. A smart rocket can come from a thousand klicks away to kill you; it doesn't care and it won't be in the least upset that it has blasted you to atomic debris. That's the real horror of modern war, that it is impersonal. Being cut with a sword hurts, and if you are close enough to do it, you can't miss the other's pain.
Steve Perry

One of man's finest qualities is described by the simple word, guts - the ability to take it. If you have the discipline to stand fast when your body wants to run, if you can control your temper and remain cheerful in the face of monotony or disappointment, you have guts, in the soldiering sense. This ability to take it must be trained-the training is hard, mental as

well as physical. But once ingrained, you can face and flail the enemy as a soldier, and enjoy the challenges of life as a civilian.
Colonel John S. Roosma

You remember the Duke of Wellington was talking of the Battle of Waterloo when he said that it was not that the British soldiers were braver than the French soldiers. It was just that they were brave five minutes longer. And in our struggles sometimes that's all it takes-to be brave five minutes longer, to try just a little harder, to not give up on ourselves when everything seems to beg for our defeat.
Paul H. Dunn

Let the soldier be abroad if he will, he can do nothing in this age. There is another personage,- a personage less imposing in the eyes of some, perhaps insignificant. The schoolmaster is abroad, and I trust to him, armed with his primer, against the soldier in full military array.
Henry Peter

No soldier starts a war-they only give their lives to it. Wars are started by you and me, by bankers and politicians, newspaper editors, clergymen who are ex-pacifists, and Congressmen with vertebrae of putty. The youngsters yelling in the streets, poor lads, are the ones who pay the price.
Francis Duffy

The glory of a workman, still more of a master workman, that he does his work well, ought to be his most precious possession; like the honor of a soldier, dearer to him than life.
Thomas Carlyle

The Lord gets his best soldiers out of the highlands of affliction.
Charles Haddon Spurgeon

Soldiers are sworn to action; they must win; some flaming, fatal climax with their lives. Soldiers are dreamers; when the guns begin, they think of homes, clean beds, and wives.
Siegfried Sassoon

You can always tell an old soldier by the inside of his holsters and cartridge boxes. The young ones carry pistols and cartridges: the old ones, grub.
George Bernard Shaw

ABRUPT, adj. Sudden, without ceremony, like the arrival of a cannon-shot and the departure of the soldier whose interests are most affected by it.
Ambrose Bierce

War loses a great deal of its romance after a soldier has seen his first battle.
Captain John S. Mosby (The Gray Ghost)

We are parlor soldiers. The rugged battle of fate, where strength is born, we shun.
Ralph Waldo Emerson

RECRUIT, n. A person distinguishable from a civilian by his uniform and from a soldier by his gait.
Ambrose Bierce

Every lover is a soldier and has his camp in Cupid.
Ovid

As an old soldier, I admit the cowardice: it's as universal as seasickness, and matters just as little.
George Bernard Shaw

In the final choice, a soldier's pack is not so heavy a burden as a prisoner's chains.
General Dwight D. Eisenhower

Stand your ground. Don't fire unless fired upon, but if they mean to have a war let it begin here!
Captain John Parker

The more you sweat in training, the less you will bleed in battle.
Motto of Navy Seals

Some warriors look fierce, but are mild. Some seem timid, but are vicious. Look beyond appearances; position yourself for the advantage.
Deng Ming-Dao

I dislike death, however, there are some things I dislike more than death. Therefore, there are times when I will not avoid danger.
Mencius

Unless you do your best, the day will come when, tired and hungry, you will halt just short of the goal you were ordered to reach, and by halting you will make useless the efforts and deaths of thousands.
General George S. Patton

Even though you hold a sword over my heart I will not give up.
Bushido Quote

And if a warrior does not manifest courage on the outside and hold enough compassion within his heart to burst his chest, he cannot become a retainer. Therefore, the monk pursues courage with the warrior as his model, and the warrior pursues the compassion of the monk.
From the Hagakure

Let me tell you what happened briefly. There were 114,000 separate aerial sorties in 42 days and one every 30 seconds. Eighty-eight thousand tons of bombs were dropped. Only seven per cent were guided. Ninety-three percent were free-falling bombs that hit where chance, necessity and no free will took them. There were 38 aircraft lost by the US in the slaughter. That number is less than the accidental losses in war games where no live ammunition is even used. No enemy aircraft rose to meet them. When the ground war came there was no ground war. Name one battle. It wasn't a battle, it was a slaughter. General Kelly said when the troops finally moved forward that there weren't many of them left alive to fight. We killed at least 125,000 soldiers and to date 130,000 civilians. We killed as many as we dared.
General Ramsey Clark

If a warrior makes loyalty and filial piety one load, and courage and compassion another, and carries these twenty-four hours a day until his shoulders wear out, he will be a samurai.
From the Hagakure

For those that will fight for it...freedom has a flavor the protected shall never know.
Lance Corporal Edwin L. Craft

From camp to camp through the foul womb of night
The hum of either army still sounds.
William Shakespeare

When the blast of war blows in our ears
Then imitate the action of the tiger,
Stiffen the sinews, summon up the blood,
Disguise fair nature with ill-favour'd rage.
William Shakespeare

O God of battles! steel my soldiers' hearts.
William Shakespeare

They come like sacrifices in their trim,
And to the fire-eyed maid of smoky war
All hot and bleeding will we offer them.
William Shakespeare

All quiet along the Potomac tonight,
No sound save the rush of the river,
While soft falls the dew on
The face of the dead
The pickets off duty forever.
Ethel Lynn Beers

In case you haven't noticed, we dehumanize our own soldiers, not because of their religion or race, but because of their low social class. Send'em anywhere. Make'em do anything. Piece of cake.
Kurt Vonnegut

The more you sweat in peace, the less you bleed in war.
Admiral Hyman G. Rickover

Weapons are an important factor in war, but not the decisive one; it is man and not materials that counts.
Mao Tse Tung

In war there is no second prize for the runner-up.
General Omar N. Bradley

The best form of welfare for the troops is first-class training.
Field Marshal Rommel

The patriot volunteer, fighting for country and his rights, makes the most reliable soldier on earth.
General Thomas J. (Stonewall) Jackson

Some people live their entire lifetime and wonder if they ever made a difference to the world. Marines don't have that problem.
Ronald Reagan

Superior firepower is an invaluable tool when entering negotiations.
General George S. Patton

Chapter Seven – Peace and War

Peace and War

BLOWIN' IN THE WIND
Bob Dylan 1963

How many roads must a man walk down
Before you call him a man?

Yes, 'n' how many seas must a white dove sail
Before she sleeps in the sand?

Yes, 'n' how many times must the cannon balls fly
Before they're forever banned?

The answer, my friend, is blowin' in the wind,
The answer is blowin' in the wind.

Man is a sentient being, aching for peace, longing for justice, praying for compassion, yet driven to conquer and vanquish. However, in war there are no answers, no truths, no peace: it simply is what it is. Only through the grace of almighty God will there be peace - and Man will not surrender quietly.

- The Editors

I believe in the doctrine of non-violence as a weapon of the weak. I believe in the doctrine of non-violence as a weapon of the strongest. I believe that a man is the strongest soldier for daring to die unarmed.
Mahatma Gandhi

The purpose of all war is peace.
Saint Augustine

Master the divine techniques of the Art of Peace and no enemy will dare to challenge you.
Ueshiba

Peace cannot be achieved through violence; it can only be attained through understanding.
Albert Einstein

PEACE, n. In international affairs, a period of cheating between two periods of fighting.
Ambrose Bierce

Nurture strength of spirit to shield you in sudden misfortune. But do not distress yourself with imaginings. Many fears are born of fatigue and loneliness. Beyond wholesome discipline, be gentle with yourself.
Desiderata

There are no fixed limits. Time does not stand still. Nothing endures, nothing is final.
Chuang Tzu

Let each one understand the meaning of sincerity and guard against display.
Chuang Tzu

Obtaining victory may be easier than preserving the results.
Joseph Lumpkin

He who joyfully marches to music in rank and file has already earned my contempt. He has been given a large brain by mistake, since for him the spinal cord would fully suffice. This disgrace to civilization should be done away with at once. Heroism at command, senseless brutality, deplorable love-of-country stance, how violently I hate all this, how despicable and ignoble war is; I would rather be torn to shreds than be a part of so base an action! It is my conviction that killing under the cloak of war is nothing but an act of murder.
Albert Einstein

In Nature, things move violently to their place and calmly in their place.
Sir Francis Bacon

When the world is peaceful, a gentleman keeps his sword by his side.
Wu Tzu

Courage is the price that life exacts for granting peace.
Amelia Earhart

What is life? It is the flash of a firefly in the night. It is the breath of the buffalo in the wintertime. It is the little shadow, which runs across the grass and loses itself in the sunset.
Crowfoot, a Blackfoot warrior

The tumult and shouting dies; the captains and kings depart; still stand the sacrifice, a humble and contrite heart. Lord God of Hosts, be with us yet, lest we forget - lest we forget.
Rudyard Kipling

The end of our Way of the Sword is to be fearless when confronting our inner enemies and our outer enemies.
Tesshu Yamaoka

But they that wait upon the LORD shall renew their strength; they shall mount up with wings as eagles; they shall run, and not be weary; and they shall walk, and not faint.
Holy Bible, Book of Isaiah

I count him braver who overcomes his desires than him who conquers his enemies: for the hardest victory is the victory over self.
Aristotle

You don't have to have fought in a war to love peace.
Geraldine A. Ferraro

Don't tell me peace has broken out.
Bertoltd Brecht

Out of a martial art, out of combat I would feel something peaceful. Something without hostility.
Bruce Lee

We seek peace, knowing that peace is the climate of freedom.
General Dwight D. Eisenhower

A brave and passionate man will kill or be killed. A brave and calm man will preserve life. Of these two types, which is good and which does harm?
Lao Tzu

A toast to the weapons of war, may they rust in peace.
Robert Orben

'Twere best at once to sink to peace, Like birds the charming serpent draws, To drop head-foremost in the jaws of vacant darkness and to cease.
Alfred, Lord Tennyson

A mind at peace, a mind centered and not focused on harming others, is stronger than any physical force in the universe.
Wayne Dyer

Americans will listen, but they do not care to read. War and peace must wait for the leisure of retirement, which never really comes: meanwhile it helps to furnish the living room.
Anthony Burgess

As peace is the end of war, so to be idle is the ultimate purpose of the busy.
Samuel Johnson

Atoms for peace. Man is still the greatest miracle and the greatest problem on this earth.
David Sarnoff

Be at war with your vices; at peace with your neighbors, and let every new year find you a better man.
Benjamin Franklin

Democracies are indeed slow to make war, but once embarked upon a martial venture are equally slow to make peace and reluctant to make a tolerable, rather than a vindictive, peace.
Reinhold Niebuhr

Disarm, disarm. The sword of murder is not the balance of justice. Blood does not wipe out dishonor, nor violence indicate possession.
Julia Ward Howe

Establishing lasting peace is the work of education; all politics can do is keep us out of war.
Maria Montessori

God is day and night, winter and summer, war and peace, surfeit and hunger.
Heraclitus of Ephesus

I don't want peace that passeth understanding, I want understanding which bringeth peace.
Helen Keller

I intend to leave after my death a large fund for the promotion of the peace idea, but I am skeptical as to its results.
Alfred Nobel

I will not by the noise of bloody wars and the dethroning of kings advance you to glory: but by the gentle ways of peace and love.
Thomas Traherne

If you cannot find peace within yourself, you will never find it anywhere else.
Marvin Gaye

Imagine all the people living life in peace. You may say I'm a dreamer, but I'm not the only one. I hope someday you'll join us, and the world will be as one.
John Lennon

In the arts of peace, man is a bungler.
George Bernard Shaw

It is easy enough to be friendly to one's friends. But to befriend the one who regards himself as your enemy is the quintessence of true religion. The other is mere business.
Mahatma Gandhi

It is wise statesmanship which suggests that in time of peace we must prepare for war, and it is no less a wise benevolence that makes preparation in the hour of peace for assuaging the ills that are sure to accompany war.
Clara Barton

Let us ever remember that our interest is in concord, not in conflict; and that our real eminence rests in the victories of peace, not those of war.
William McKinley

Life is pleasant. Death is peaceful. It's the transition that's troublesome.
Isaac Asimov

Mankind must remember that peace is not God's gift to his creatures; peace is our gift to each other.
Elie Wiesel

The people of the world genuinely want peace. Some day the leaders of the world are going to have to give in and give it to them.
General Dwight D. Eisenhower

Nothing contributes more to a person's peace of mind than having no opinions at all.
G. C. Lichtenberg

Obsessed by a fairy tale, we spend our lives searching for a magic door and a lost kingdom of peace.
Eugene O'Neill

He does not want war but if war is to be fought it is to be won. True victory is in finding a path that does not lead to conflict. Conflict leads to a lesser victory, which is based on survival and not resolution. Ultimate victory is the saving of all life and honor. This is winning without fighting.
Joseph Lumpkin

The arts of peace and war are like two wheels of a cart, which, lacking one, will have difficulty in standing.
Kuroda Nagamasa

Higher worth is like water. Water is good at benefiting ten thousand beings without vying for position... In dwelling, be close to the land. In heart and mind value depth. In interacting with others, value kindness. In words, value reliability. In rectifying, value order. In social affairs, value ability. In action, value timing. In general, simply don't fight and hence have no blame.
Lao Tzu

Life is a hard battle anyway. If we laugh and sing a little as we fight the good fight of freedom, it makes it all go easier. I will not allow my life's light to be determined by the darkness around me.
Sojourner Truth

An unjust peace is better than a just war.
Marcus Tullius Cicero

Whoever undertakes to set himself up as a judge of Truth and Knowledge is shipwrecked by the laughter of the Gods.
Albert Einstein

Victory goes to the player who makes the next-to-the-last mistake.
Savielly Grigorievitch Tartakower

In order to make an apple pie from scratch, you must first create the universe.
Carl Sagan

A designer knows he has achieved perfection not when there is nothing left to add, but when there is nothing left to take away.
Antoine de Saint-Exupery

A man's feet should be planted in his country, but his eyes should survey the world.
George Santayana

Every gun that is made, every warship launched, every rocket fired signifies, in the final sense, a theft from those who hunger and are not fed, those who are cold and are not clothed.
General Dwight D. Eisenhower

A society grows great when old men plant trees whose shade they know they shall never sit in.
Greek Proverb

An eye for an eye makes the whole world blind.
Mahatma Gandhi

When I hear somebody sigh, life is hard, I am always tempted to ask, compared to what?
Sydney J. Harris

In times like these, it is helpful to remember that there have always been times like these.
Paul Harvey

The loud little handful will shout for war. The pulpit will warily and cautiously protest at first. The great mass of the nation will rub its sleepy eyes, and will try to make out why there should be a war, and they will say earnestly and indignantly: It is unjust and dishonorable and there is no need for war. Then the few will shout even louder. Before long you will see a curious thing: anti-war speakers will be stoned from the platform, and free speech will be strangled by hordes of furious men who still agree with the speakers but dare not admit it...Next, statesmen will invent cheap lies, putting blame upon the nation that is attacked, and every man will be glad of those conscience-soothing falsities, and will diligently study them, and refuse to examine any refutations of them; and thus he will by and by convince himself that the war is just, and will thank God for the better sleep he enjoys after this process of grotesque self-deception.
Mark Twain

I do not want the best to be any more the deadly enemy of the good. We climb through degrees of comparison.
Archbishop Edward White Benson

The only way to win World War III is to prevent it.
General Dwight D. Eisenhower

To be prepared for war is one of the most effectual means of preserving peace.
George Washington

The final battle against intolerance is to be fought - not in the chambers of any legislature - but in the hearts of men.
General Dwight D. Eisenhower

To map out a course of action and follow it to an end requires some of the same courage that a soldier needs. Peace has its victories, but it takes brave men and women to win them.
Ralph Waldo Emerson

Soldier rest! thy warfare o'er, Sleep the sleep that knows not breaking, Dream of battled fields no more, days of danger, nights of waking.
Sir Walter Scott

It is a brave act of valor to condemn death, but where life is more terrible than death it is then the truest valor to dare to live.
Sir Thomas Brown

Wars are not acts of God. They are caused by man, by man-made institutions, by the way in which man has organized his society. What man has made, man can change.
Frederick Moore Vinson

Great spirits have always found violent opposition from mediocre minds. The latter cannot understand it when a man does not thoughtlessly submit to hereditary prejudices but honestly and courageously uses his intelligence.
Albert Einstein

Since wars begin in the minds of men, it is in the minds of men that the defenses of peace must be constructed.
UNESCO Constitution

Peace is more important than all justice; and peace was not made for the sake of justice, but justice for the sake of peace.
Martin Luther

Bravery and cowardice are not things that can be conjectured in times of peace. They are in different categories.
From the Hagakure

Liberty is never unalienable; it must be redeemed regularly with the blood of patriots or it always vanishes. Of all the so-called natural human rights that have ever been invented, liberty is the least to be cheap and is never free of cost.
Robert A. Heinlein

Four score and seven years ago our fathers brought forth on this continent a new nation, conceived in liberty, and dedicated to the proposition that all men are created equal.
Abraham Lincoln

War does not determine who is right - only who is left.
Anonymous

War would end if the dead could return.
Stanley Baldwin

Weapons are tools of violence; all decent men detest them.
Lao Tzu

What difference does it make to the dead, the orphans and the homeless, whether the mad destruction is wrought under the name of totalitarianism or in the holy name of liberty and democracy?
Mahatma Gandhi

I am not only a pacifist but a militant pacifist. I am willing to fight for peace. Nothing will end war unless the people themselves refuse to go to war.
Albert Einstein

They wrote in the old days that it is sweet and fitting to die for one's country. But in modern war, there is nothing sweet nor fitting in your dying. You will die like a dog for no good reason.
Ernest Hemingway

Strike against war, for without you no battles can be fought! Strike against manufacturing shrapnel and gas bombs and all other tools of murder! Strike against preparedness that means death and misery to millions of human beings! Be not dumb, obedient slaves in an army of destruction! Be heroes in an army of construction!
Helen Keller

Man is the only animal that deals in that atrocity of atrocities, War. He is the only one that gathers his brethren about him and goes forth in cold blood and calm pulse to exterminate his kind. He is the only animal that for sordid wages will march out and help to slaughter strangers of his own species who have done him no harm and with whom he has no quarrel. And in the intervals between campaigns he washes the blood off his hands and works for the universal brotherhood of man with his mouth.
Mark Twain

We do not admire a man of timid peace.
Theodore Roosevelt

The soldier, above all other people, prays for peace, for he must suffer and bear the deepest wounds and scars of war.
General Douglas MacArthur

What the horrors of war are, no one can imagine. They are not wounds and blood and fever, spotted and low, or dysentery, chronic and acute, cold and heat and famine. They are intoxication, drunken brutality, demoralization and disorder on the part of the inferior ... jealousies, meanness, indifference, selfish brutality on the part of the superior.
Florence Nightingale

To give victory to the right, not bloody bullets, but peaceful ballots only, are necessary.
Abraham Lincoln

It is only those who have neither fired a shot nor heard the shrieks and groans of the wounded who cry aloud for blood, more vengeance, more desolation. War is hell.
William Tecumseh Sherman

Humanize war? You might as well talk about humanizing hell!
British Admiral Jacky Fisher

You can have peace, or you can have freedom. Don't ever count on having both at once.
Robert A. Heinlein

Let us recollect that peace or war will not always be left to our option; that however moderate or un-ambitious we may be, we cannot count upon the moderation, or hope to extinguish the ambition of others.
Alexander Hamilton

We make war that we may live in peace.
Aristotle

Where there is no peril in the fight, there is no glory in the triumph.
Pierre Corneille

In peace sons bury fathers, but war violates the order of nature, and fathers bury sons.
Heroditus

I don't know whether war is an interlude during peace, or peace is an interlude during war.
Georges Clemenceau

The most disadvantageous peace is better than the most just war.
Desiderius Erasmus

It is war that shapes peace, and armament that shapes war.
Thomas Fuller

You can no more win a war than you can win an earthquake.
Jeannette Rankin

War may sometimes be a necessary evil. But no matter how necessary, it is always an evil, never a good. We will not learn how to live together in peace by killing each other's children.
Jimmy Carter

War is not its own end, except in some catastrophic slide into absolute damnation. It's peace that's wanted. Some better peace than the one you started with.
Lois McMaster Bujold

The only winner in the War of 1812 was Tchaikovsky.
Solomon Short

Wars teach us not to love our enemies, but to hate our allies.
W. L. George

You can't separate peace from freedom because no one can be at peace unless he has his freedom.
Malcolm X

Never believe that a few caring people can't change the world. For indeed, that's all who ever have.
Margaret Mead

Do not fear your enemies, the worst they can do is kill you. Do not fear your friends, at worst, they may betray you. Fear those who do not care, they neither kill nor betray, but betrayal and murder exist because of their silent consent.
Bruno Jasienski

You can't say that civilization don't advance, for in every war they kill you in a new way.
Will Rogers

No more wars, no more bloodshed. Peace unto you. Shalom, salaam, forever.
Menachem Begin

Peace hath her victories, no less renowned than war.
John Milton

As peace is the end of war, it is the end, likewise, of preparations for war; and he may be justly hunted down, as the enemy of mankind, that can choose to snatch, by violence and bloodshed, what gentler means can equally obtain.
Samuel Johnson

Peace and justice are two sides of the same coin.
General Dwight D. Eisenhower

With malice toward none, with charity for all, with firmness in the right, as God gives us to see the right, let us strive on to finish the work we are in, to bind up the nation's wounds.
Abraham Lincoln

Lamb of God, who taketh away the sins of the world, have mercy on us;
Lamb of God, who taketh away the sins of the world, have mercy on us;
Lamb of God, who taketh away the sins of the world, grant us peace.
Agnus Dei

Epilogue

War does not end when treaties are signed. The dead still rot in the fields; images of their dying burned into our minds. Now is the time to bury them and turn a blind eye; or to try…to try. The earth bloats with blood that ran. Life goes on. We have crops to plant, and we eat the bread grown in soil once fed by the blood of brothers.

"The tumult and shouting dies; the captains and kings depart; still stand the sacrifice, a humble and contrite heart. Lord God of Hosts, be with us yet, lest we forget - lest we forget. "
   Rudyard Kipling

Now the War is in our souls.

~ The Editors

Behold the Second Horseman